GENETICS: A BASIC GUIDE

GENETICS

A BASIC GUIDE

I. J. PEDDER

&

E. G. WYNNE

W · W · NORTON & COMPANY · INC · NEW YORK

Library of Congress Cataloging in Publication Data
Pedder, I J
Genetics: a basic guide.
1. Genetics. I. Wynne, E. G., joint author.
II. Title. [DNLM: 1. Breeding. 2. Genetics.
SF105 L616g 1958]
QH430.P42 1974 575.1 73-21553
ISBN 0-393-09267-4

Printed in the United States of America

1 2 3 4 5 6 7 8 9 0

Contents

Illustrations

PLATES

FIGURES

Acknowledgements

During the writing of this text we have received invaluable advice, criticism and assistance from a large number of people. In particular we would like to record our gratitude to the following:

Professor A. J. Cain, of Liverpool University, who read and criticised the section on *Cepaea*. Professor A. D. Bradshaw, of Liverpool University, who supplied information on heavy metal tolerance in plants. Mr B. J. F. Haller, of Harris Biological Supplies, and Dr D. I. Southern, of Manchester University, who have been most generous in their help with photographic material. Dr G. E. Anderson, of Liverpool Polytechnic, for allowing us to use some of her experimental methods with *Coprinus lagopus*. Mrs Valerie Hall, who typed the manuscript with the utmost care and patience. Finally to our wives, who have provided us with a constant source of help and encouragement.

The following have kindly allowed us to reproduce diagrams, data and photographs:

The Literary Executor of the late Sir Ronald A. Fisher, F.R.S., to Dr Frank Yates, F.R.S., and to Oliver and Boyd, Edinburgh, for permission to reprint Table 3 from *Statistical Tables for Biological, Agricultural and Medical Research* (Sixth edition, 1963). Figure 13: By permission of *Unilever Educational Publications*. Figure 60: By permission of *Biological Sciences Curriculum Study*. Figure 77: By permission of The Royal Society, and the authors. Tables 6 and 7: By permission of Dr H. B. D. Kettlewell, the University of Oxford. Figure 39: By permission of Dr M. Wallace and the editor of *School Science Review*. Figure 29: By permission of Dr W. Crowther and Macmillan. Figures 74 and 75 and Plate X: By permission of Professor A. D. Bradshaw. Figures 81 and 82 (a), (b): From C. O. Carter, *Human Heredity*, Penguin (1962). Table 4: From *Variation and Evolution in Plants*, by G. L. Stebbins, Columbia University Press (1950). Table 2 and Figure 8: From *Genetics* by R. P. Levine. Copyright © 1962 by Holt, Rhinehart and Winston. Reprinted by permission of Holt, Rhinehart and Winston. Plate XIV: From *Introduction to Fungi* by Professor J. Webster, reproduced by permission of Cambridge University Press.

I

The genetic material

1.1. INTRODUCTION

ONE of the fundamental properties of all living things is their ability to reproduce themselves from their own body materials. The necessity of such a process is clear, for the life of any single organism is finite and, if the species as a whole is to continue, individuals must produce new individuals during their lifetime. In the nineteenth century Louis Pasteur showed that even bacteria were only produced by a process of reproduction of existing bacteria. Until that time it was widely held that many organisms arose from inorganic material by the process of spontaneous generation. Thanks to the work of Pasteur and others it is evident that all living organisms arise from pre-existing organisms and, what is more, from organisms of basically the same type. The principle of 'like begets like' is almost too obvious to be worth stating, but, on reflection, it is remarkable that a species reproduces its basic plan so faithfully generation after generation. A closer examination of the individuals within a species, however, reveals considerable variation imposed upon the overall basic plan. Such variation within a species is almost always on a comparatively superficial level and yet, as we shall see later, may be of fundamental importance both to the individual and the species. The science of genetics is concerned with the mechanisms by which the basic plan of the organism, and its variations, are transmitted from parent to offspring during the process of reproduction. The genetic mechanisms will depend, partially, on the mode of reproduction which is used by the animal or plant and thus we must consider the two main types of reproduction and their consequences for the organism.

1.2. ASEXUAL REPRODUCTION

Organisms are said to reproduce asexually where offspring are produced by a process of separation from the parental body without the fusion of special reproductive cells from two parents. Such reproduction is common in many plants and lower animals. One of the simplest examples is the process of fission which takes place in Protozoa. In

Amoeba the parent cell divides into two more or less identical daughter cells, a process termed *binary fission*. Occasionally there is repeated division within the parent cell of the nucleus before the cytoplasm divides to give a number of new organisms. This process, termed *multiple fission* or *sporulation*, is seen in the sporozoans, for example *Plasmodium*. Asexual reproduction by fragmentation is seen in some aquatic annelids such as *Lumbriculus*, flatworms and ribbon worms. Coelenterates also show this method of reproduction in, for example, the transverse division of the scyphistoma of jelly-fish and the fragmentation of sea anemones from their base. Budding is a type of reproduction where a new individual arises as an outgrowth of an older animal and is characteristic of Coelenterates. The bud grows and develops while attached to the parent and may then separate, as in *Hydra*, or remain attached to give rise to a colony of many individuals such as *Obelia*. Asexual reproduction in plants takes two forms. Firstly asexually produced spores such as in mosses, liverworts and ferns where the spores link the sporophyte and gametophyte generations. Spore production is also seen when there is no alternation of gametophyte and sporophyte in the life history as a means of rapid dissemination of the organism. In the algae examples include *Chlamydomonas* and *Vaucheria*, and in the fungi *Mucor, Pythium, Eurotium* and *Erysiphe*. The second method is by the separation from the parent of a large, well-differentiated organ of propagation. Examples include the gemmae of liverworts, stem and root tubers, corms, bulbs, runners, suckers and stolons of flowering plants. This type of reproduction is seen at its most specialised in the production of entire plantlets around the margin of the leaf in *Bryophyllum*. The chief importance of asexual reproduction, as far as genetic mechanisms are concerned, is that because the offspring are derived directly from the cells of a single parent they will show virtually the same characteristics as the parent. This statement must be qualified by saying that parent and offspring will only be identical if they live in identical environments.

1.2.1. *Sexual reproduction*

Most animals and plants reproduce by a process involving the production of specialised sex cells termed *gametes*. Typically there are two types of gamete, male and female, each of which is generally incapable of giving rise to a new organism itself. Fusion of a male with a female gamete, the process of *syngamy* or *fertilisation*, produces the *zygote*, a unicellular structure which will give rise to a new individual under the correct conditions.

The two types of gametes may be produced on the same individual, when it is termed *hermaphrodite*, or *bisexual*, or on separate individuals of the same species in which case it is *unisexual*. From a genetic standpoint the vital result of sexual reproduction, as opposed to asexual, is that the offspring show variation both from their parents and between themselves, even when grown under identical conditions.

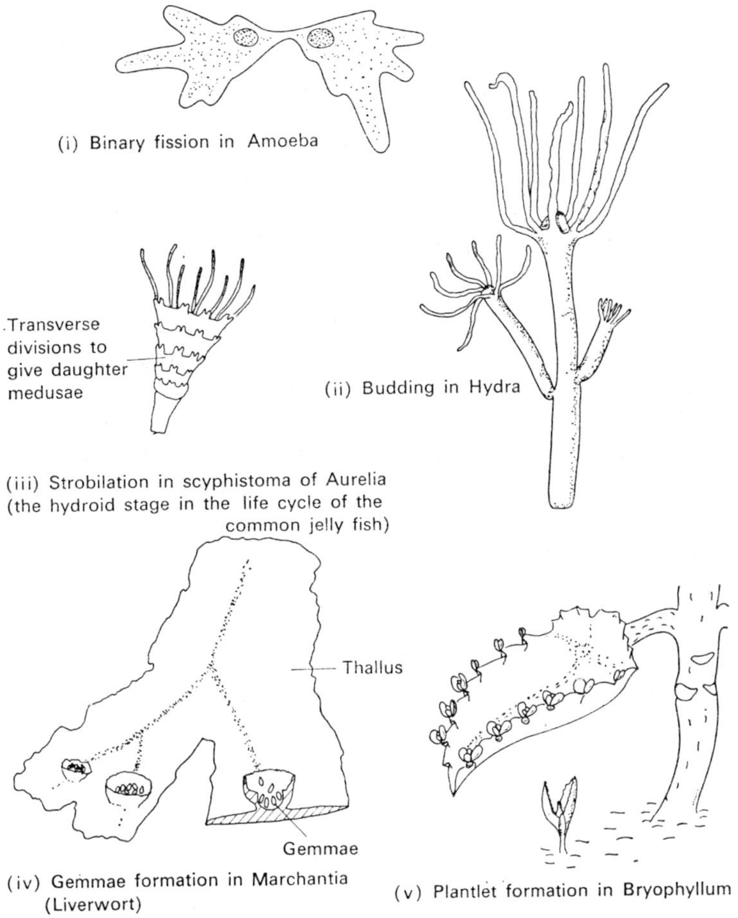

FIG. 1
Various forms of asexual reproduction

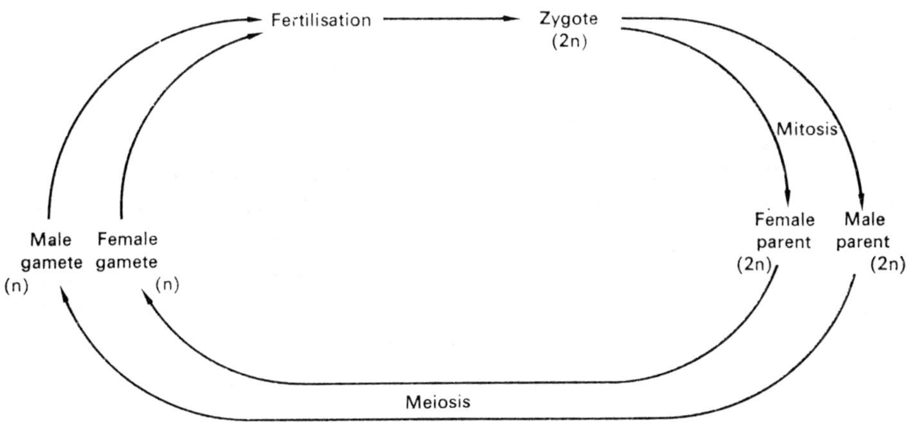

FIG. 2
Sexual reproduction

1.3. THE IMPORTANCE OF THE GAMETES

From the generalised scheme of sexual reproduction shown earlier it is clear that the gametes provide the only link between the parents and the offspring. Therefore any information passed from parents to offspring must be carried within these cells. Such information will include not only the basic plan of the organism but also the variation which exists between individuals over and above this basic plan. In the vast majority of organisms the gametes are microscopic, the eggs of birds being a notable exception. It was not until towards the end of the seventeenth century that de Graaf showed that the eggs of mammals were produced in the ovaries and Leeuwenhoek and Hamm used the microscope to show the presence of what are now termed sperms in seminal fluid. The activity of these millions of sperms convinced some people at the time that it was the male element that was important in transmitting hereditary information. One observer even went so far as to claim that he could see a miniature human figure (which he termed a homunculus) 'preformed' inside a sperm. Another school of thought adopted the opposite theory and stated that it was the egg cell that transmitted all the hereditary information, the sperm was merely an 'activator'. Evidence that both sperm and eggs were required to produce a new individual was obtained in 1785 by Spallanzani who filtered out the sperms from the seminal fluid of frogs and toads and showed that no fertilisation and embryo formation took place when this seminal fluid was mixed with eggs. In 1875 Hertwig observed the process of fusion of gametes in sea urchins and noted that it involved the fusion of the sperm and egg nuclei. A similar gradual understanding of the importance of the gametes in plants can be seen. In 1694 Cammerarius showed that corn plants produced no seed unless pollen was applied to the carpels. In the eighteenth century Linneaus and Koelreuter carried out many experiments on the hybridisation of plants. They showed that when pollen of one species is placed on the stigma of another, any offspring combine the characteristics of both species, thus indicating that both pollen and ovules carry hereditary information.

Although the morphology and size of gametes varies tremendously between species, all gametes have certain essential features in common. Wherever they are clearly defined they are unicellular structures, each with a nucleus. When a species produces gametes which are, in appearance, identical it is said to be *isogamous*, such as in some algae, fungi and protozoa. It should be stated that in most cases of isogamy there are physiological differences between the gametes. Most organisms are *anisogamous*, that is they produce two distinctly different types of gametes, male and female. In general the female gamete is non-motile and larger than the male gametes which are usually provided with specialised organelles which enable them to move to the female gamete. A notable exception is found in flowering plants where the male gamete, formed within the pollen grain, is non-motile. The formation of the pollen tube, however, from the pollen grain provides a means whereby the male gamete is conducted directly to the female gamete in the ovule.

The fusion of two gametes during the process of syngamy results in the formation of a zygote, the essential process of sexual reproduction. The zygote cell has, within

FIG. 3
Mammalian gametes

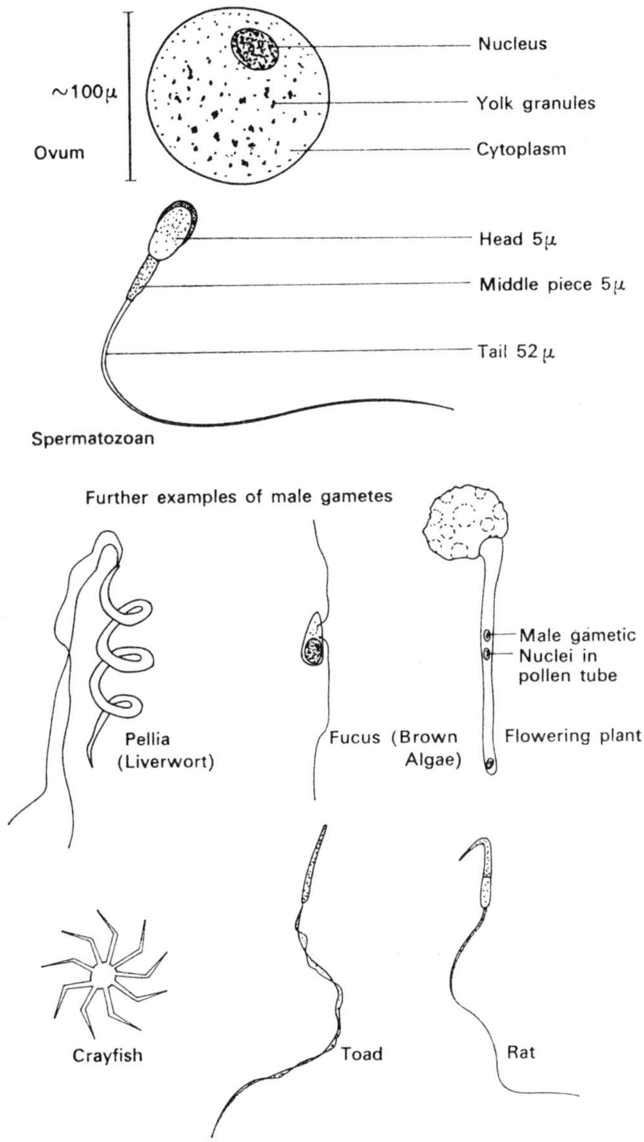

itself, all the information which is necessary to form a new individual organism. The most important feature of syngamy is that it involves the fusion of the nuclei of the gametes, one male gamete nucleus with one female gamete nucleus, indeed the majority of male gametes seem to consist of little else but a nucleus. After the penetration of one male gamete into the female gamete, changes in the chemical nature of the outer region of the female gamete prevent the entry of further male gametes. In the few cases where further male gametes do enter, they are not involved in nuclear fusion. The precise mode of entry of the male gamete, its movement within the female gamete and the timing of nuclear fusion all vary from species to species, but in every case there is a direct fusion of the nuclei of the two gametes. Early observers of the process of fertilisation, such as Strasburger and Hertwig in 1884, claimed that children displayed charac-

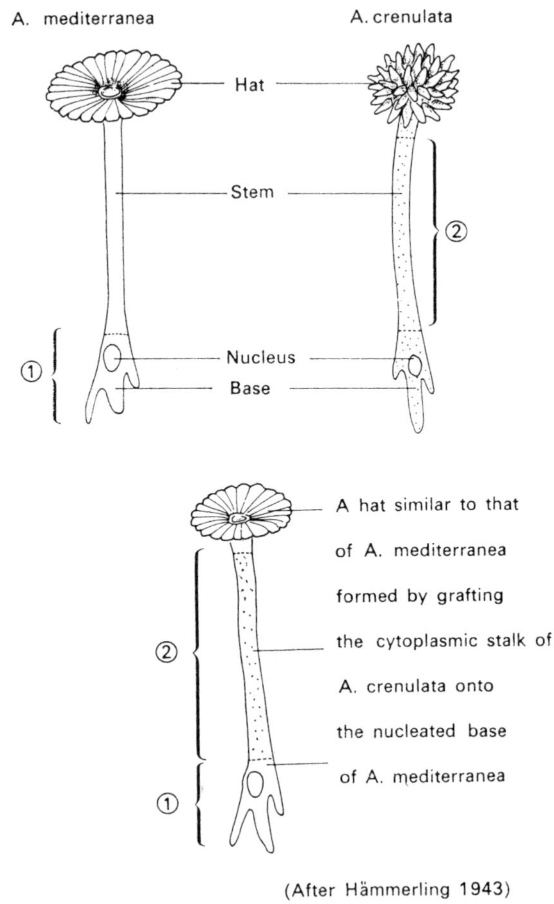

FIG. 4
Grafting experiments in *Acetabularia*

ters from both parents, to a more or less equal extent, and therefore both gametes must make equal contributions to the offspring. It has previously been pointed out that there are quite significant morphological and physiological differences between the two types of gamete in most species. Such things as size, the amount of cytoplasm, the presence of locomotory structures are among the more obvious points. Clearly then if both gametes are to make equal contributions to the offspring we must look for some structure which is more or less equivalent in the two types as the vehicle carrying the hereditary information.

There is much evidence which implicates the nucleus as being responsible for controlling the development of the cell, but the single example of *Acetabularia* will suffice. *Acetabularia* is a unicellular alga with three main parts: a holdfast, a stalk and a cap. The single nucleus is in the holdfast. If the stalk and the cap are removed the nucleated holdfast will regenerate the stalk and cap typical of that species. If a nucleated holdfast is grafted on to the stalk of another species of *Acetabularia*, with quite a distinct cap, the cap which is eventually formed will be of the type one would expect in the species from which the nucleated holdfast was taken. Clearly, then, the nucleus is capable of controlling the formation of the cap, even through the cytoplasm of the stalk of a different species.

1.4. THE STRUCTURE OF THE NUCLEUS AND THE NUCLEAR CONTENTS

The development of the electron microscope, mainly over the last quarter of a century, has led to a spectacular increase in our knowledge of cell structure. Using a beam of electrons, instead of light, and electromagnetic fields to focus the beam, objects as small as 10 Å can be clearly resolved (1 Å $= 10^{-8}$ cm). Plate I is a photograph taken through the electron microscope of onion-root tip cells, and shows clearly the degree of detail of cell structures visible at this magnification. Figure 5 shows the structures within an animal cell at a comparable magnification. The use of the electron microscope has revealed that the nucleus consists of a nuclear membrane of double unit membrane structure, which is connected to the endoplasmic reticulum. In fact the nuclear membrane may be regarded as part of the endoplasmic reticulum which has come to surround the genetic material. It is also clear from many electron microscope preparations that the nuclear membrane is penetrated by a large number of pores, possibly up to 10% of its total surface area. The contents of the nucleus include an apparently structureless nucleoplasm which seems to consist largely of proteins, thin, lightly staining threads, the chromosomes and one or more densely staining areas, the nucleoli.

In a normal, non-dividing cell the chromosomes are generally indistinguishable from the nucleoplasm. During cell division they undergo various changes, for instance shortening and coiling, which render them more visible. They are seen as darkly staining rods, variable in length and the vast majority with a constriction, the *centromere*, at some point along their length. In fact, chromosomes are generally classified according to the position of their centromere. For any species of living organism it is possible to describe the chromosomes of any cell, for, with one exception, all the cells of every individual within a species have an identical number of chromosomes of, from the point of view of appearance, identical chromosomes. This 'set' of chromosomes of a particular species is referred to as the *karyotype* of the species. The karyotypes of different species vary greatly in number: *Pisum sativum*, the garden pea, has 14 chromosomes, *Nicotiana tabaccum*, the tobacco, has 48, some ferns have up to 400 chromosomes. In the animal kingdom the nematode worm *Ascaris* has 2 chromosomes, man has 46, some radio-

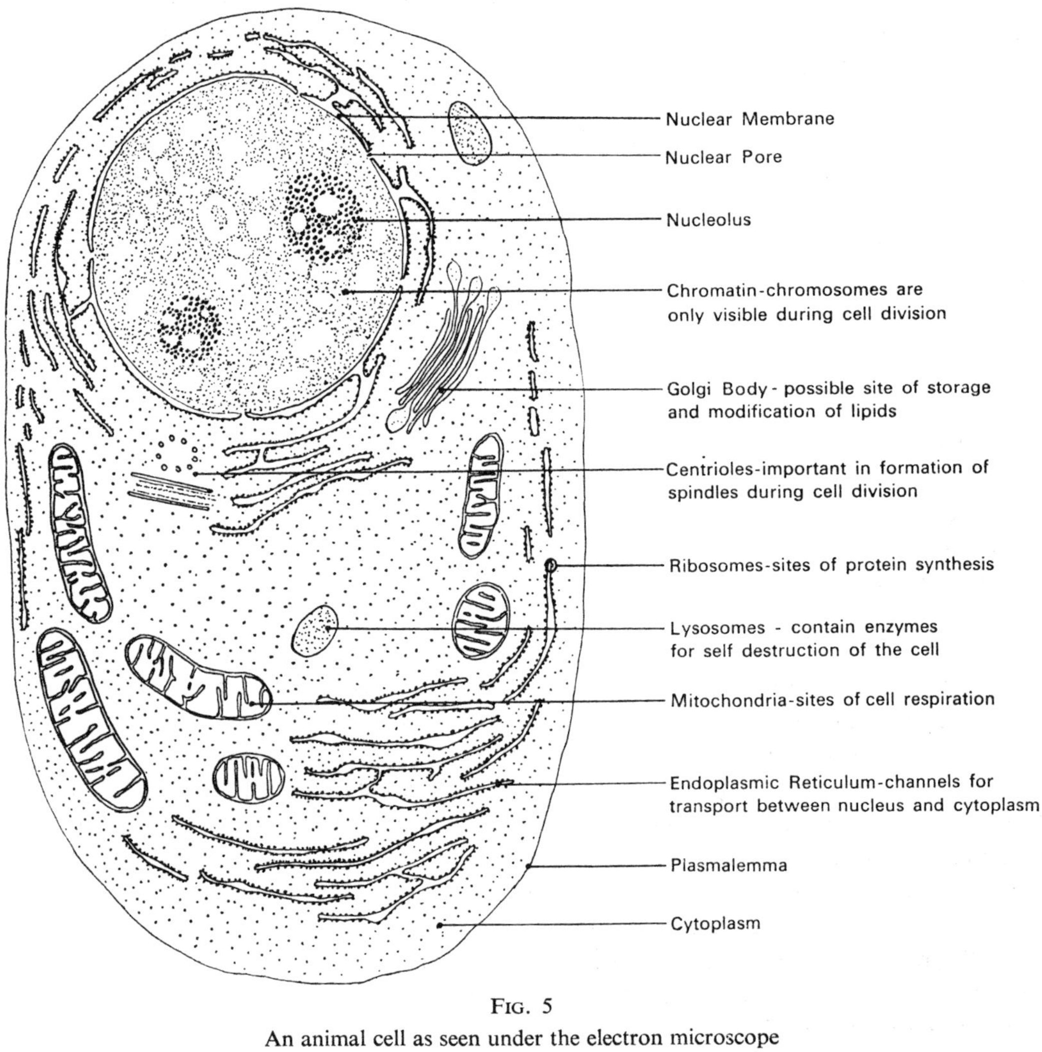

FIG. 5
An animal cell as seen under the electron microscope

larians (protozoa) have 1600. Generally speaking animals seem to have more chromosomes, but they are usually smaller, ranging from 4 to 6 μ (1 $\mu = 10^{-3}$ mm) in length, whereas some plant chromosomes reach 50 μ. The single exception to the constancy of the karyotype is found in the gametes, which are found to have only half the number of chromosomes of all the other cells. The significance of this vital fact will be explained in detail at a later stage. The nucleolus seems to be connected with a particular chromosome and in some cases is made up of two distinct regions, a fibrous element the nucleoneme towards the outside, and the more amorphous pars ammorpha internally.

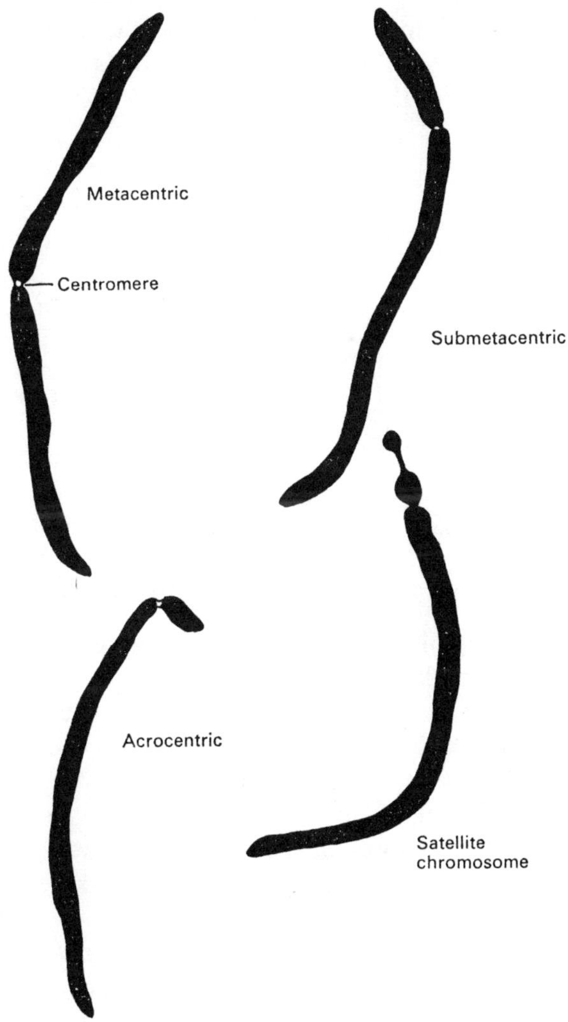

FIG. 6
The different types of chromosomes

1.5. THE FINE STRUCTURE OF CHROMOSOMES

The chromosomes are coiled structures, consisting of one or more threads, the *chromonemata*, coiled together. The degree of coiling of the chromosome determines the thickness and therefore the visibility of the chromosome; during cell division they are tightly coiled, short, thick and clearly visible. When the cell is not dividing they are loosely coiled, long, thin and therefore not visible. Each chromonemata consists of a number of *fine fibrils* down to about 50 Å in diameter and each having basically two types of molecule; protein and deoxyribonucleic acid (DNA). It is not known for certain how these fibrils are built up into the chromonemata, the most plausible suggestion is that they consist of protein 'blocks' joined by strands of DNA, and that these are able to coil upon themselves to form the chromonemata.

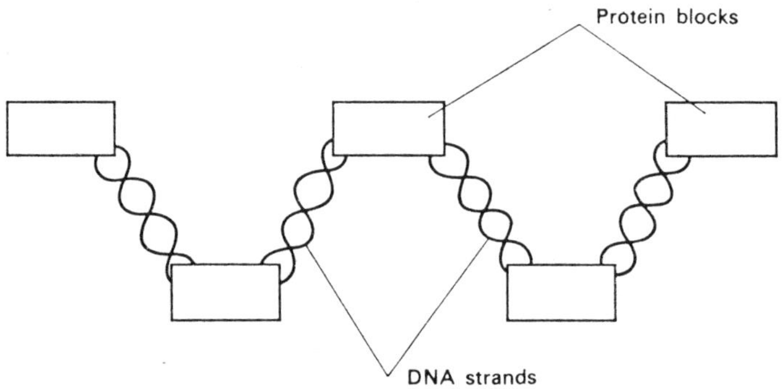

FIG. 7
Suggested structure of chromosome fine fibrils

1.6. EVIDENCE THAT DNA IS THE HEREDITARY MATERIAL

There are a number of strong pieces of evidence to suggest that the hereditary information is carried in the DNA molecule.

1. Circumstantial evidence is provided by the fact that DNA is almost always associated with structures which seem to be involved in genetic mechanisms, such as chromosomes and, as we shall see later, mitochondria.

2. It is to be expected that, if the transmission of the genetic material is carried out by a particular chemical compound, then that compound should be metabolically stable. The DNA molecule has been found to be very stable compared with carbohydrates and proteins for instance.

3. It has been possible, using several methods, to extract the DNA from cells and determine how much DNA there is present in each nucleus. Chemical extraction and purification of the DNA from a known number of cells is one such method, and a second involves the use of the Feulgen reaction. This reaction produces a reddish-

purple pigmentation with DNA and the intensity of the reaction can be measured by sensitive photometric equipment, thus indicating the amount of DNA present. Both methods indicate that in a given species all the body cells have the same amount of DNA in their nuclei, while the gametes have half that amount. Such constancy

TABLE 1. Amount of DNA in the nucleus of various types of cell in chickens.

Cell	*DNA in mg* $\times$ *10*$^{-9}$
Red blood cell	2·49
Liver cell	2·66
Heart cell	2·45
Kidney cell	2·20
Pancreas cell	2·61
Spleen cell	2·55
Sperm cell	1·26

of DNA content is to be expected if the DNA is the genetic material responsible for controlling the functioning of all cells.

4. Direct evidence that the DNA contains the genetic material has come from the work on bacteria involving transforming principles. In the 1920's F. Griffith, working on the bacterium *Diplococcus pneumoniae* (pneumococcus), found that it could exist as two different types: a smooth form (*S*) in which the bacterial cell is covered with a polysaccharide coating and which is virulent, and a rough form (*R*) which has no coating and is non-virulent. The two types of bacteria reproduce to form their own type, so the characteristics are genetically controlled. In addition there is a variety of types of *S* pneumococcus, termed types *IIS* and *IIIS*, depending on the exact constitution of their polysaccharide coating, and this characteristic is also inherited for generation after generation. Griffith injected animals with a mixture of living non-virulent *R* bacteria and heat-killed type *IIIS* bacteria, neither of which was capable of causing pneumonia on its own. He found, however, that some animals died of pneumonia and in their blood he found, in addition to *R* bacteria, live, virulent type *IIIS* bacteria. Thus, some of the live *R* bacteria had been transformed into type *IIIS* bacteria, presumably by the dead type *IIIS* bacteria injected with them. Later workers repeated the experiments *in vitro* and found that an extract of the type *IIIS* bacteria was capable of transforming the *R* bacteria. The chemical nature of this substance, the transforming principle, was discovered in 1944 by Avery, MacCleod and McCarty who found that highly purified extracts of DNA from the type *IIIS* bacteria were able to transform the *R* bacteria into type *IIIS*. If the DNA was previously treated with deoxyribonuclease, an enzyme that breaks up the DNA molecule, no transformation takes place. It is now known that such transformation can take place in a number of species of bacteria,

and it is clear that other large molecules, such as proteins which are present in the bacterial cells, are not able to effect transformation.

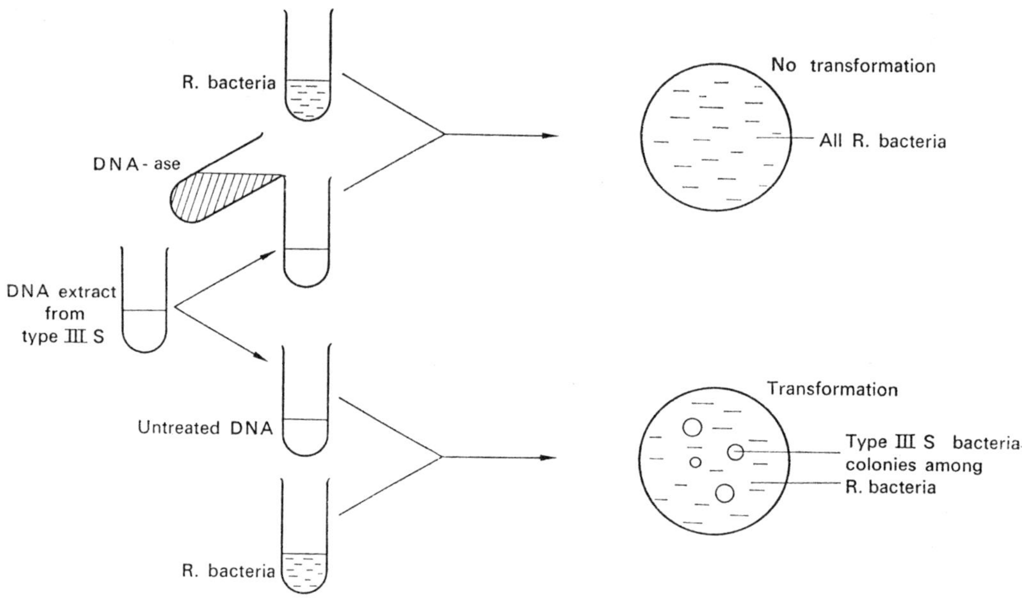

FIG. 8
Summary of transformation in pneumococcus

5. Further implication of DNA as the genetic material has come from the study of the reproduction of bacterial viruses, or bacteriophages. The structure of a bacteriophage particle is shown in Fig. 9 and it is seen to consist of head and tail regions. The outer layers of the particle are proteinaceous while the head region contains DNA. During infection of a bacterial cell the tail of the bacteriophage becomes attached to the outside of the cell. The wall of the bacterium is pierced, and material is injected from the phage into the cell. After a short while the infected cell bursts open, a process termed lysis, and hundreds of phage progeny are released. The progeny, which are formed by the utilisation of bacterial protein and DNA forming mechanisms, have the same genetic characteristics as the original phage particle. It has been shown, by using phage particles which have either their DNA content or their protein content radioactively labelled, that it is the phage DNA which is injected into the bacterial cell, and which controls the reproduction of the genetically identical phage progeny. Experiments on these lines were performed in 1952 by Kershey and Chase. They grew host bacteria in a medium containing either radioactive sulphur (S^{35}) or phosphorus (P^{32}). These isotopes were incorporated by the bacteria, which thus became 'labelled'. These bacteria were infected with phage, which reproduced and their progeny were found to

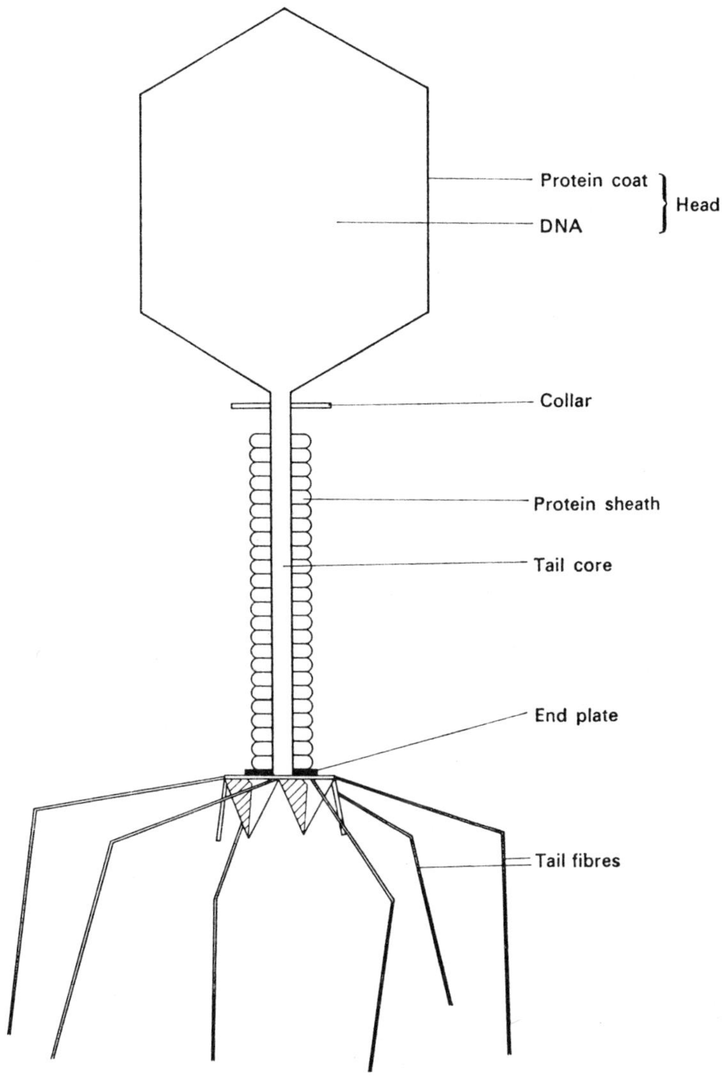

FIG. 9

The structure of a T bacteriophage

From *The Genetics of Bacterial Virus* R. S. Edgar and R. H. Epstein. Copyright © (February 1965) by *Scientific American*.

be labelled with S^{35} or P^{32} according to which bacteria they infected. The vital point is that in those phage which were labelled with S^{35} it was the protein which was labelled, as DNA does not contain sulphur, and in those labelled with P^{32} it was the DNA which was labelled since the protein does not contain phosphorus. These labelled phage were then used to infect unlabelled bacteria and the distribution of the label inside the bacterial cells noted after infection. With the phage labelled with S^{35} the bacterial

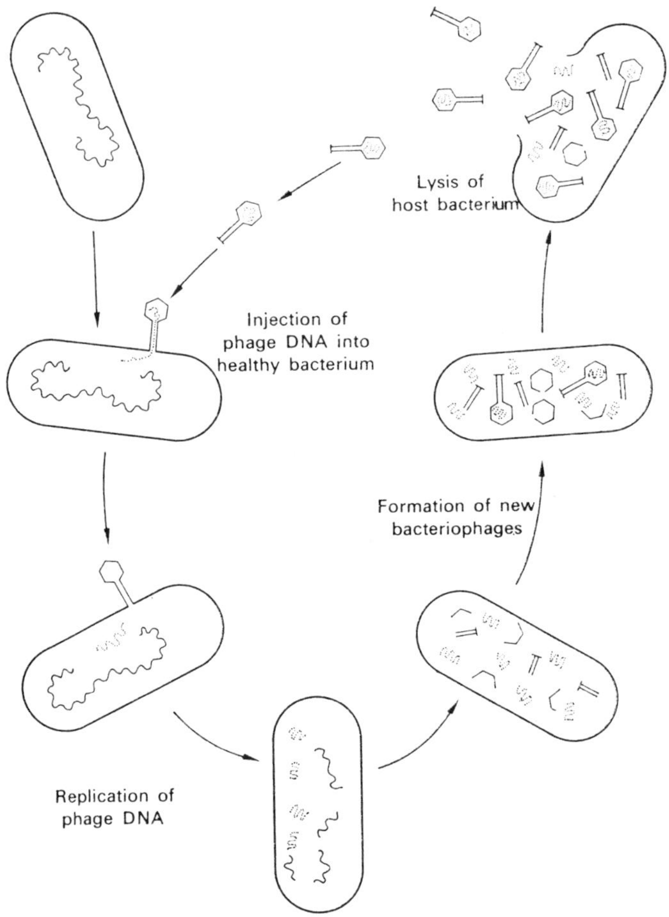

FIG. 10

The life history of a bacteriophage

From *Viruses and Genes*, François Jacob and Elie L. Wollman. Copyright © (June 1961) by *Scientific American*.

protoplasm remained unlabelled, but the labelled protein coat of the phage was found outside the bacterial cells. In the case of the P^{32} labelled phage, the protoplasm of the bacterial cells they infected was found to be labelled, indicating that the DNA of the phage had entered the bacterial cell to initiate the production of further phage particles.

1.7. CYTOPLASMIC INHERITANCE

By considering a wide range of information we have come to the conclusion that the hereditary information is transmitted from generation to generation within the substance DNA, and that this substance is carried on the chromosomes lying in the nuclei

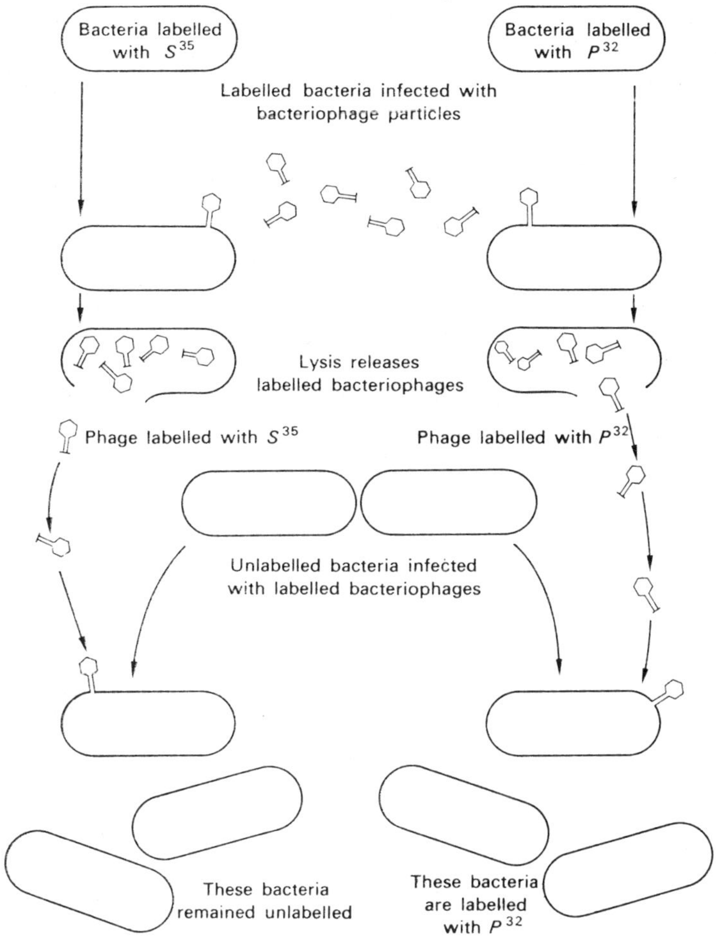

FIG. 11
Plan of the experiment to show that DNA carries the heredity information using labelled bacteriophages

of the gametes. It would be wrong, however, to end this chapter with the assumption that this was the only mechanism by which information is passed to the progeny. There are a small number of cases where it is clear that the cytoplasm, or cytoplasmic inclusions, exert an influence on the hereditary mechanisms, a fact which is less surprising when one realises that the vast majority of cell functions are carried out in the cytoplasm, even though they may be controlled by nuclear DNA.

1.7.1. *Maternal effects*

In a case where the cytoplasm carried hereditary information it would seem likely that the contribution of such information made by the male and female parents would not be equal, since the female gamete almost always has a larger quantity of cytoplasm. In such a case one would expect differences in the progeny between reciprocal crosses, i.e. crosses between two types of organism where the source of male and female gametes is reversed. An example studied by Rhoades is that of 'male sterility' in maize, in which affected plants produce little or no pollen. A cross between a 'male sterile' female and a normal male produced almost all 'male sterile' progeny. The reciprocal cross of a normal female and a 'male sterile' male (a very small quantity of pollen produced) gave progeny which were all normal. Further tests by Rhoades showed that the important factor in determining this character in the offspring was contained in the maternal cytoplasm.

A striking case of a maternal effect is seen in the pond snail *Limnaea peregra.* In this species both types of coiling of the shell are found, those which coil to the right (dextral) and those which coil to the left (sinistral). The character of coiling is not,

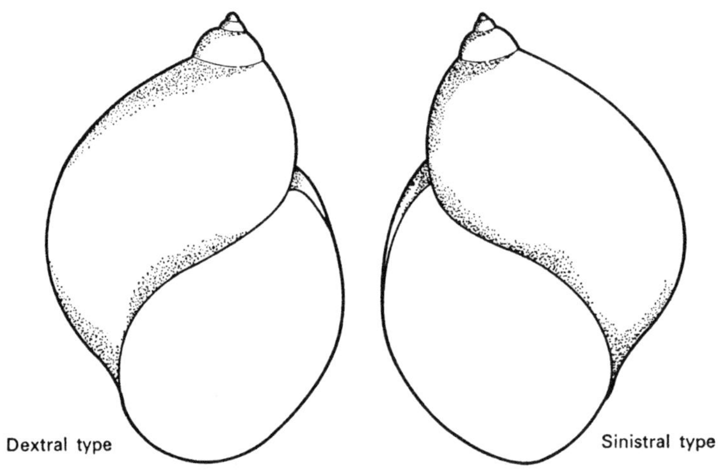

FIG. 12
Coiling in *Limnaea* shells

however, determined by its own DNA, but by that of its mother. The maternal DNA impresses a certain pattern on the cytoplasm of the female gametes which controls the early development of the zygote. It is during these early stages of development that the eventual direction of coiling of the shell is laid down, and therefore this character is determined by the maternal DNA but transmitted via the cytoplasm of the female gametes.

1.7.2. *Cytoplasmic inclusions*

A certain strain of mice have a very high incidence of a type of mammary cancer, and reciprocal crosses showed that this cancer susceptibility was transmitted maternally. Furthermore, it was shown that the substance which induces cancer susceptibility is transmitted via the mothers' milk, hence it is termed the 'milk factor'. Recent evidence suggests that this 'milk factor' is in fact a virus, which is passed on to progeny in the milk and which induces susceptibility to cancer.

In the fruitfly, *Drosophila melanogaster*, there is a strain which differs from normal flies in that it is highly sensitive to carbon dioxide. Reciprocal crosses indicate that the majority of the progeny are the same as their mothers for this character, thus indicating that the trait was carried in the cytoplasm of the female gametes. Further investigation showed that virus-like particles could be extracted from the cytoplasm of sensitive flies, and these were termed sigma particles. These sigma particles can be used to induce sensitivity in non-sensitive eggs by implantation of normal ovaries into sensitive females whereupon sigma particles are transmitted to the normal eggs in the transplanted ovaries.

1.7.3. *Cytoplasmic organelles*

Within the cytoplasm there are a large number of organelles, the presence of which is quite normal, and which are in fact vital to the functioning of the cell. In many cases the inheritance of these organelles seems to be 'extra-nuclear'. For example it is known that the matrix of mitochondria contains DNA, probably enough to control the formation of further mitochondria. It is thought that the majority of mitochondria arise by fission of existing mitochondria; this has been supported by experiments on the fungus *Neurospora* in which mitochondria were radioactively labelled. It is of great significance also that while the female gamete contains much cytoplasm in which many mitochondria may be carried, male gametes generally have very little cytoplasm and yet most male gametes carry mitochondria.

In plant cells the green pigment chlorophyll is carried in discrete structures, chloroplasts. These are developed, in most cases, from plastid primordia which, in higher plants, are transmitted by the female parent via the cytoplasm of the embryo sac. The number of primordia transmitted is small and therefore the vast majority of chloroplasts in the adult plant must arise from these few plastid primordia, or be built up individually. Information on the inheritance of chloroplasts has come from a study of variegated plants, and in particular *Mirabilis jalapa*, variegated four o'clock. It can be seen from Table 2 that the type of chloroplasts possessed by the progeny depends upon those of the female parent. Where the female parent has normal green chloroplasts the offspring always has this type of chloroplast, regardless of the male parent.

A similar situation is seen where the female parent has the abnormal pale chloroplasts. Where the female parent originates from a variegated branch the plastid primordia included in the embryo sac may be normal green, abnormal pale or both, thus giving three types of possible offspring. Thus the inheritance of chloroplasts appears to be maintained within the organelles themselves and to be independent of nuclear DNA.

TABLE 2. Chloroplast inheritance in variegated four o'clock (*after Levine*)

Branch of origin of male parent	*Branch of origin of female parent*	*Progeny*
Green	Green Pale Variegated	Green Pale Green, Pale Variegated
Pale	Green Pale Variegated	Green Pale Green, Pale Variegated
Variegated	Green Pale Variegated	Green Pale Green, Pale Variegated

2

The action of the genetic material

2.1. THE PHENOTYPE

WE have defined genetics as the study of the mechanisms by which an organism inherits its basic body plan plus a degree of variation from its parents. To recognise such mechanisms we must be conscious of the various characteristics by which the organism displays its basic plan and its variability. For example, when comparing a number of mice we may see that they are similar in respect of such characters as the possession of a vertebral column, in their dentition, in the structure of their limbs and in being warm-blooded—in fact in all the characters which 'make' them mice. On the other hand they may differ from one another in such characters as adult weight, pigmentation of their coat, reproductive capacity and food preferences. In a similar way a group of oak trees may be alike in the shape of their leaves, the distribution of vascular tissue and the production of acorns, but may differ in size, rate of water uptake and growth, and density of branching. The sum total of such characters displayed by an organism is referred to as the *phenotype* of the organism, and variation in these characters as *phenotypic variation*. The more obvious characters of the phenotype are the morphological ones, which are directly visible to the observer and which generally concern the size, shape, colour and texture of the organism. Characters which are less readily observed but, none the less, are just as important parts of the phenotype are the multitude of biophysical and biochemical reactions which go on in even the simplest organisms. A large proportion of phenotypic characters are passed on from generation to generation, but they cannot be passed on as such, except in those methods of asexual reproduction which involve a splitting off of what was originally part of the parental body, such as budding, binary and multiple fission and vegetative propagation. We have evidence that it is, in fact, the DNA contained within the gametes which transmits the hereditary information. The genetic information, in the form of DNA, which an organism obtains from its parents is referred to as the organism's *genotype*. In some way the information carried in the genotype has to be translated into the phenotypic characters which we are able to observe and measure.

2.2. THE IMPORTANCE OF ENZYMES

In order to establish a link between the genotype and the phenotype of an organism we must digress slightly at this point to a brief consideration of enzymes. The organism's phenotype is essentially a product of the vast complex of chemical reactions which go on within the cells, i.e. its metabolism. Some of these reactions which spring to mind include respiration, photosynthesis, the breakdown of waste products, the synthesis of pigments, proteins and a whole range of organic compounds. All these reactions, and many more of course, play some part in producing phenotypic characteristics. For even the simplest of metabolic processes to take place under the conditions of temperature, pressure, acidity and alkalinity (pH) which the cells of living organisms provide, some form of catalysis is essential. Enzymes are the catalysts of living organisms, the substances which enable these reactions to proceed, often at remarkable speed, under the conditions prevalent within living cells. But, apart from this ability, the most important factor of enzyme action is its specificity, whereby each enzyme is only able to catalyse at most a very restricted range of chemical changes. Some enzymes, e.g. urease, are able to catalyse only one reaction, namely the hydrolysis of urea. In some cases the degree of specificity is such that an enzyme is able to discriminate between a given compound and its optical isomer, for instance l-amino acid oxidase will not act on d-amino acids and *vice versa.* If these two properties of enzymes are combined with the fact that they are extremely active in very small quantities, it can be seen that a cell possesses the ability to regulate very closely the complex metabolic processes taking place within itself. The presence of a particular enzyme will mean that a single, or small group of, reactions will take place with precise results on the metabolic pathways and hence the phenotype.

2.3. THE PROTEIN NATURE OF ENZYMES

In 1926 Sumner obtained pure crystalline urease, the first enzyme to be isolated. On analysis the crystals of urease proved to be protein, and in recent years a large number of enzymes have been purified and all of them have been found to contain protein. Further evidence to support this is given by treating enzymes with proteinases, enzymes which break down proteins, when enzyme activity is destroyed. It is also possible to demonstrate that agents which cause denaturation of proteins, for example high temperature and a wide range of pH, also cause a parallel loss of enzyme activity. Having established that enzymes are essentially proteins it is now necessary to examine the structure of the protein molecule. Proteins are large molecules which show a great diversity of form and function, and yet which all have the same basic structure. They are chain molecules made up of a large number of individual units, amino acids. The amino acids, of which there are approximately twenty-two different types occurring

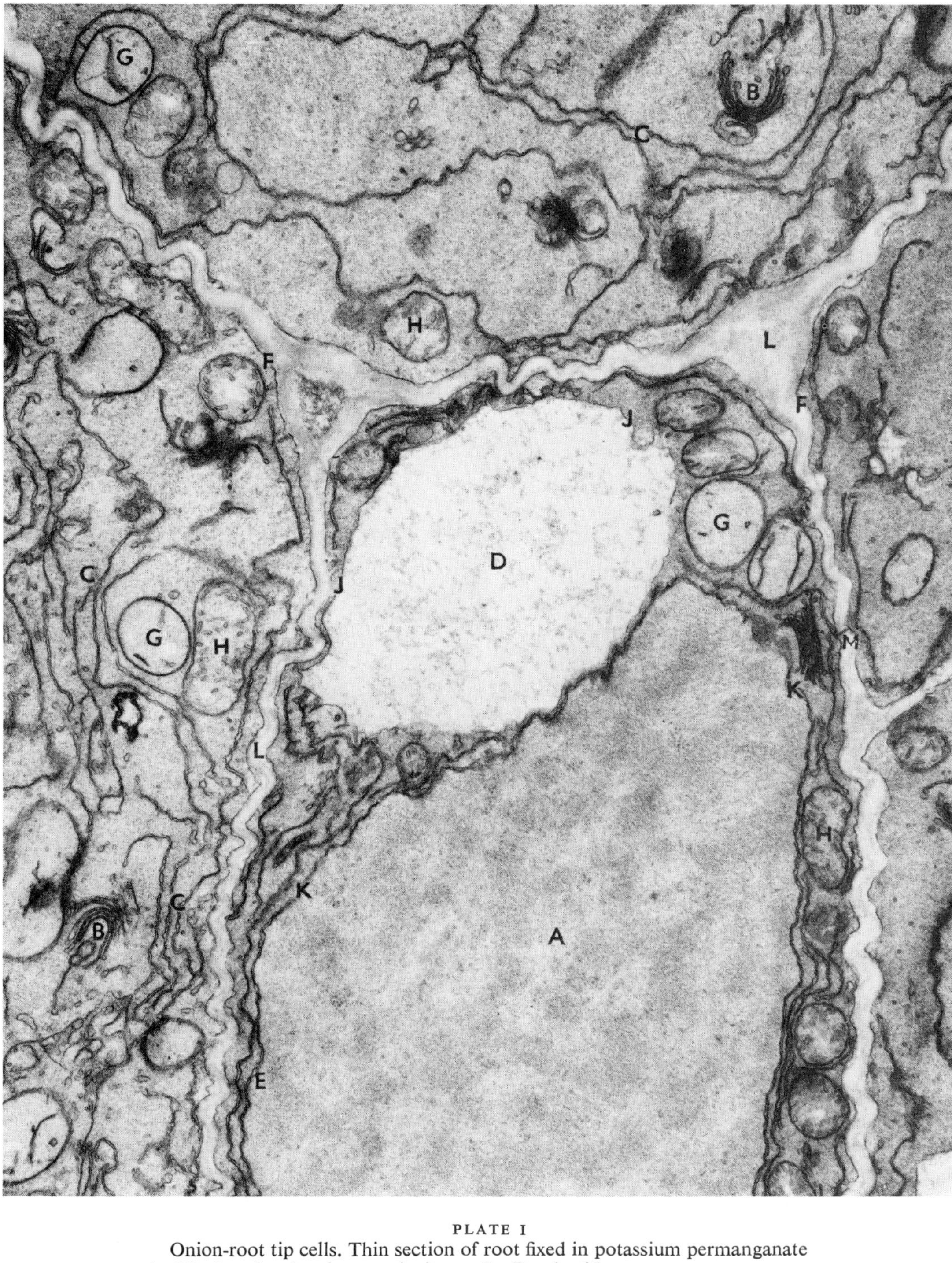

PLATE I

Onion-root tip cells. Thin section of root fixed in potassium permanganate

A	Nucleus (nucleoplasm region)	G	Proplastid
B	Dictyosome (Golgi apparatus)	H	Mitochondrion
C	Endoplasmic reticulum	J	Tonoplast (vacuolar membrane)
D	Vacuole	K	Nuclear envelope (double membrane)
E	Nuclear pore	L	Cell wall
F	Plasmalemma (cell membrane, plasma membrane)	M	Plasmodesma (pl. -ata)

a
Interphase

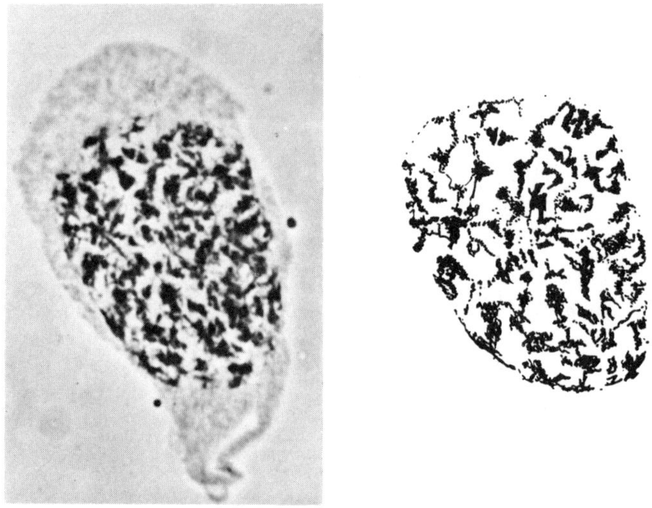

The somatic cell prior to mitosis, the nucleus is in the interphase condition

b
Early Prophase

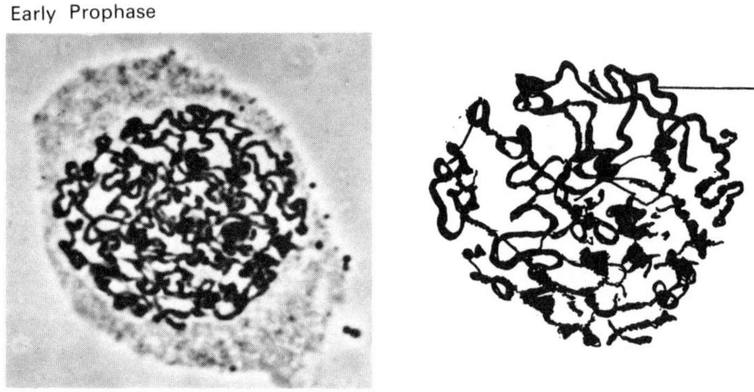

Chromosomes beginning to condense out

c
Prophase

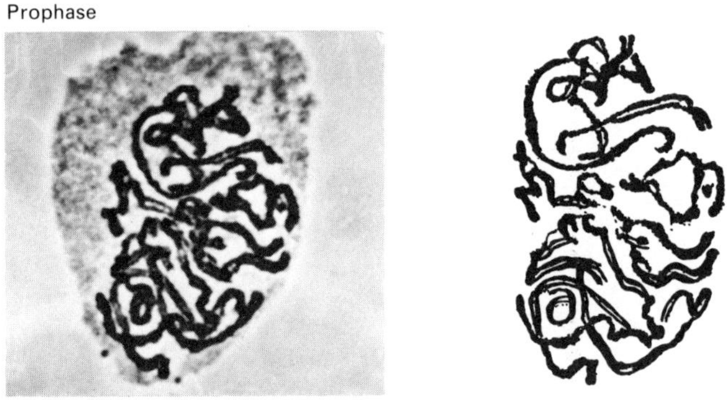

The chromosomes are more condensed, each chromosome consists of two strands or sister chromatids

PLATE II a–c
Stages in Mitosis

d
Metaphase

(Preparation treated with colchicine see chapter 6·6)

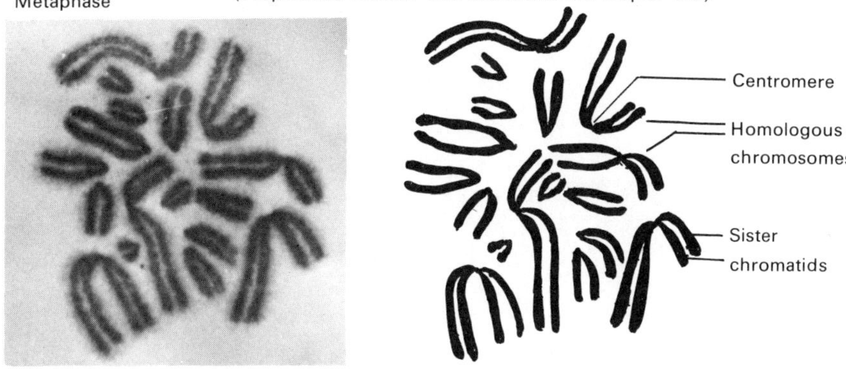

e
Mid Anaphase

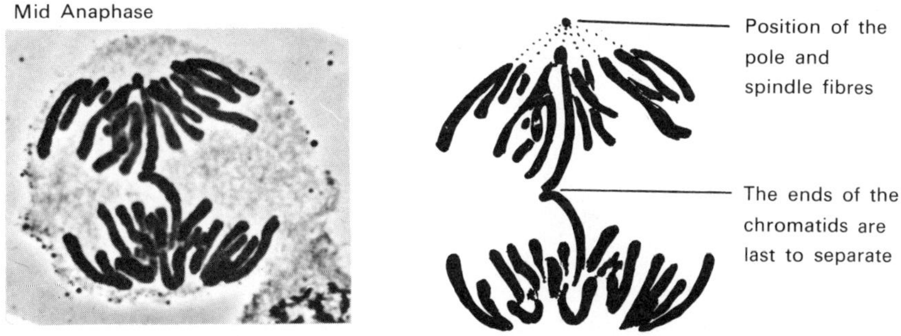

f
Early Telophase

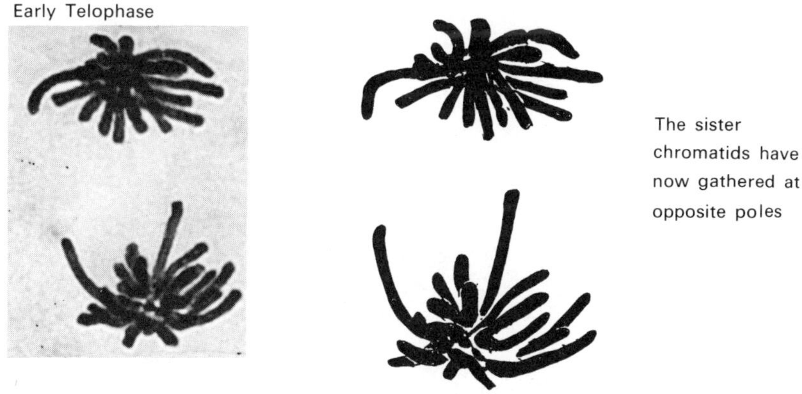

PLATE II d–f
Stages in Mitosis

g
Telophase

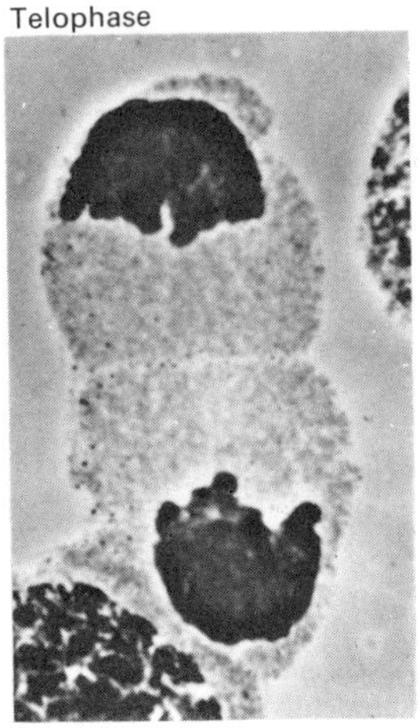

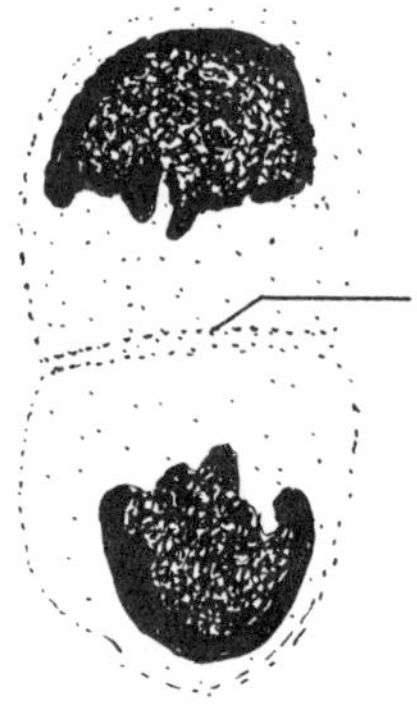

h
Interphase

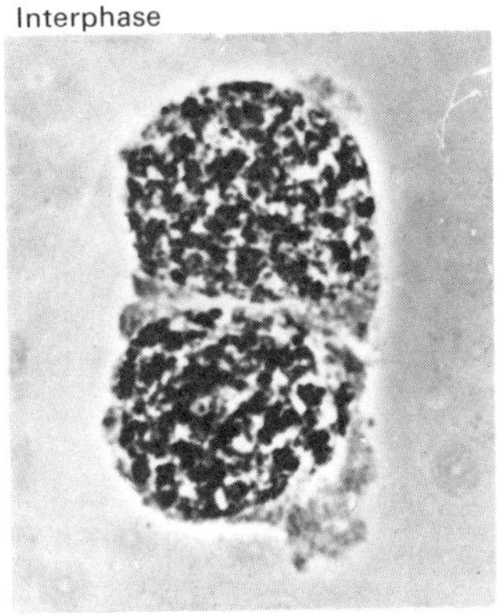

PLATE II g–h
Stages in Mitosis

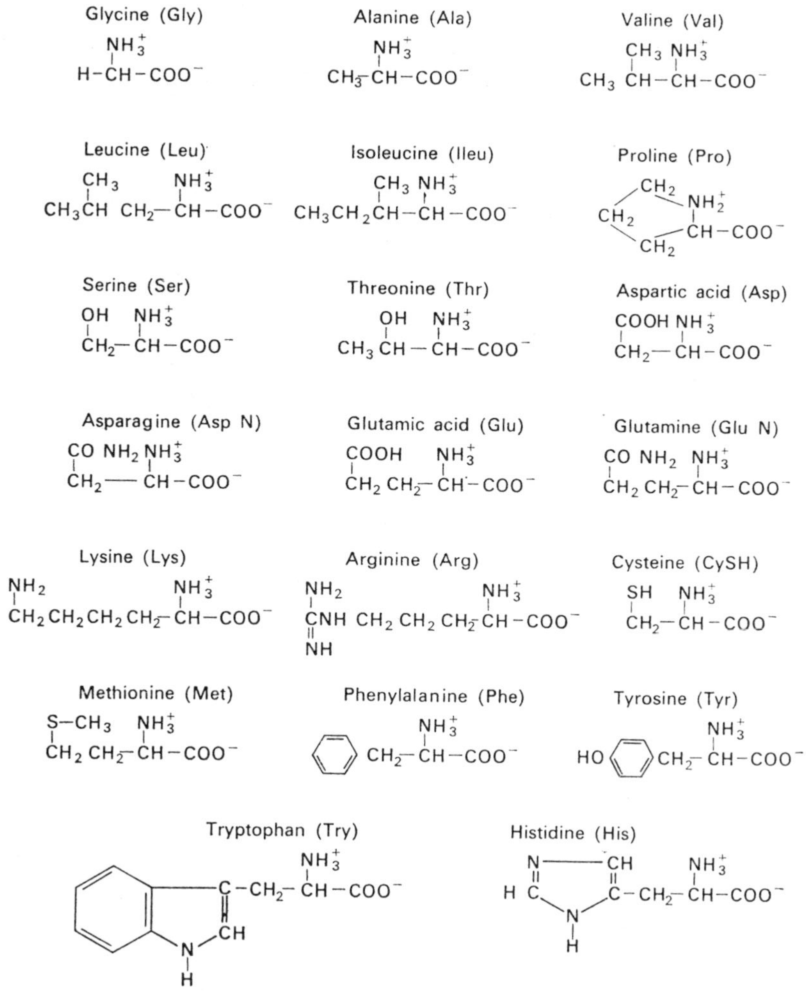

Fig. 13
The twenty common amino acids occurring in proteins

naturally, are linked together by a bond known as the peptide bond. The number and sequence of the amino acids in the protein molecule confer upon the protein its *primary structure*, just as the number and sequence of letters in a sentence confer the meaning and function of the sentence. Clearly the number of proteins which it is possible to create from the 20 amino acids is very large indeed. In some proteins there is more than one chain of amino acids, for example insulin has two such chains linked together. Two further aspects of protein structure are of great importance, namely that the protein molecule is not a long, straight chain of amino acids, but the whole molecule is thrown into a coil-like structure termed a *helix*. The helical form is the most common *secondary structure* of proteins. Furthermore, mainly as a result of reactions between the side-chains of the amino acids and water, the protein molecule will fold in a highly specific and characteristic way when in water. This process of folding confers upon the molecule a *tertiary structure*, which is characteristic for every protein. Thus, the basic structure of the protein molecule depends upon the number and sequence of amino acids, but superimposed upon and dependent upon this primary structure will be characteristic secondary and tertiary structures. The importance of the secondary and tertiary structures lies in the fact that they are almost certainly responsible for the more specific properties of the

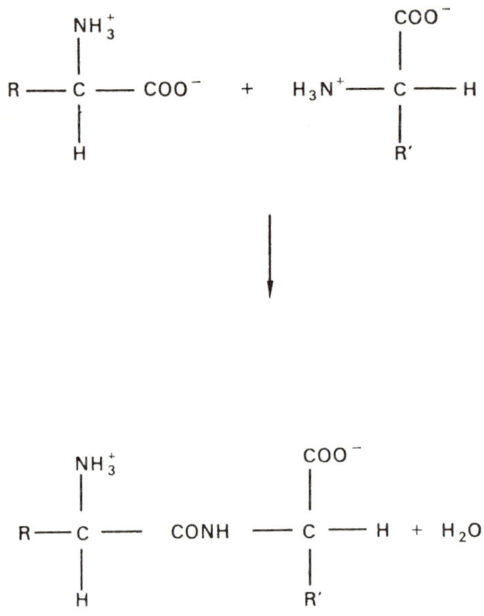

FIG. 14
Formation of a peptide bond

proteins which chiefly interest us, namely enzymes. An enzyme first combines with the molecules involved in the reaction which it is catalysing giving rise to a complex, 'enzyme + reactants'. The combination of enzyme and reactant molecules has been likened to a lock and key mechanism, the reactants fitting only into a specific part of the surface of a particular enzyme called the active centre. After the chemical transformation of the reactants has taken place they no longer fit into the active centre of the enzyme, and are released. Clearly the specificity of an enzyme will depend upon the precise folding of the protein molecule which gives it its tertiary structure, but this will in turn ultimately depend upon the primary structure of the protein as represented by the sequence of amino acids. The vital question which must now be answered is how the genetic information, in the form of nuclear DNA, affects and controls the production of the specific proteins, the enzymes, which give rise to the phenotype.

2.4. THE STRUCTURE OF DNA

Chemical analysis of DNA has shown it to consist of a deoxyribose sugar fraction, a phosphate group and four types of organic bases, adenine and guanine which are both

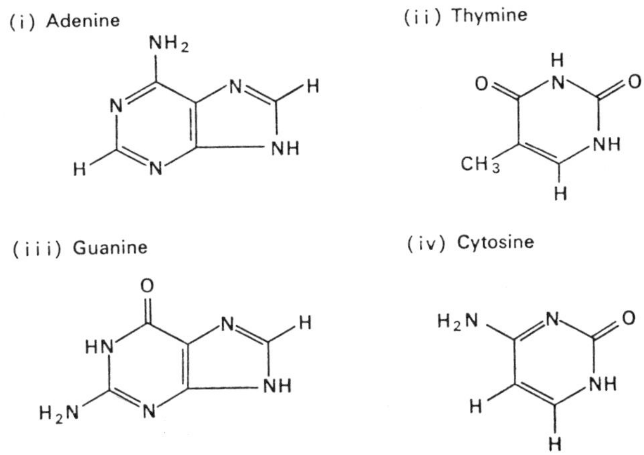

FIG. 15
The four bases present in the DNA molecule

purines and cytosine and thymine, which are both pyrimidines. Like proteins the DNA molecule is extremely large and built up of a number of smaller units, termed *nucleotides.*

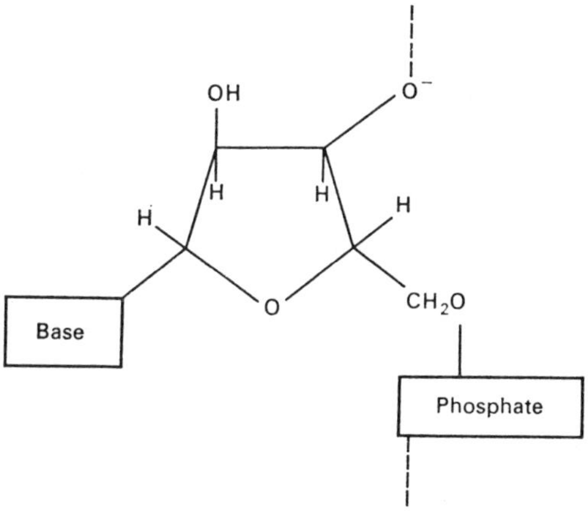

FIG. 16
A single nucleotide

Each nucleotide consists of one of the four possible organic bases linked to the sugar deoxyribose, which in turn is linked to the phosphate group. The nucleotides are linked together to form a polynucleotide chain, in which the sugar of one nucleotide is joined to the phosphate group of the next, and so on.

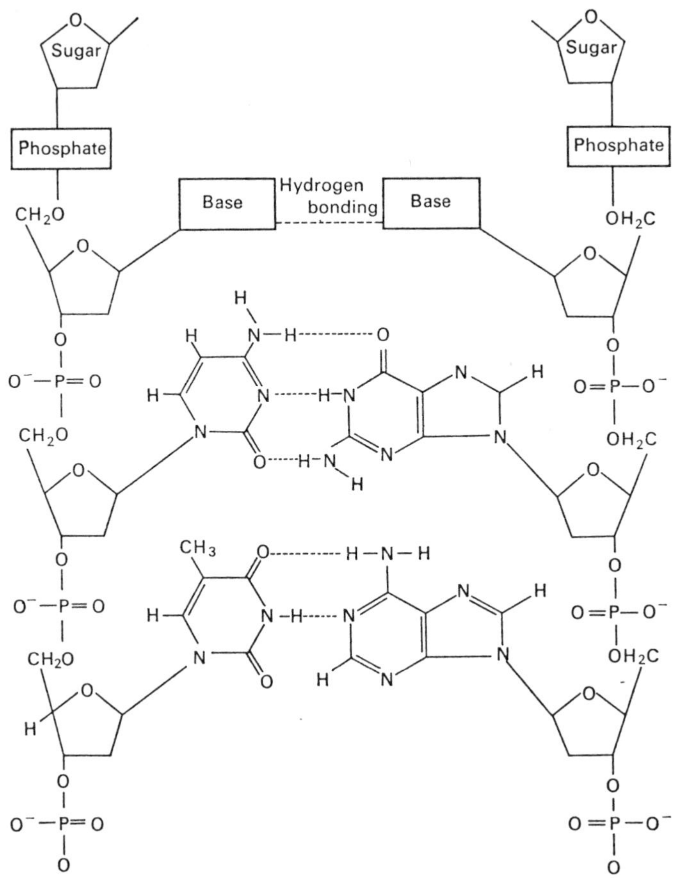

FIG. 17

The structure of, and bonding between, two polynucleotide chains

Chargaff, in 1950, showed that in any sample of DNA the amount of adenine present always equalled the amount of thymine, and similarly the amount of guanine always equalled the amount of cytosine. Thus the ratio of *A*:*T* and *G*:*C* is always 1:1, but the ratio of *AT*:*GC* varies in different organisms. At the same time Wilkins, Franklin and others were using complex techniques involving X-ray diffraction patterns to investigate the three-dimensional structure of the molecule. They showed firstly that the nucleotides were built up into the DNA macromolecule in the same way in the

DNA of a variety of organisms, and secondly that the molecule had a helical structure. In 1953 Watson and Crick suggested a structure for the DNA molecule which has been substantiated by a number of subsequent investigations. They suggested that the molecule in fact consisted of two DNA strands coiled around each other to form a double helix, and that these two strands were joined together through the organic

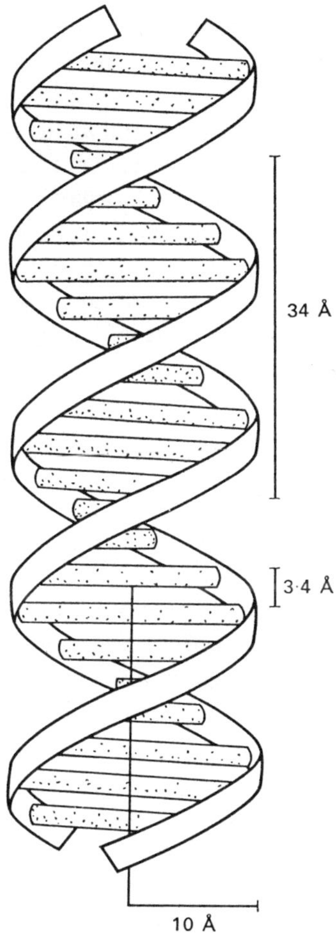

FIG. 18
Plan of the DNA helix

bases. They used Chargaff's results coupled with the fact that a pair of nucleotides can be joined by hydrogen bonds (*H* bonds), via the organic bases. Moreover, such bonding between the four organic bases in DNA is highly specific. A nucleotide with adenine forms two *H* bonds with a nucleotide with thymine and a nucleotide with guanine forms three *H* bonds with a nucleotide containing cytosine. Formation of *H* bonds between any other pairs of bases is physically very difficult and would require distortion of the helix. Thus the model of DNA proposed by Watson and Crick gave a sound explanation for the equal ratios of adenine to thymine and cytosine to guanine found by Chargaff. It can be seen that the sequence of the organic bases along one of the polynucleotide strands will automatically determine the sequence on the other strand, by virtue of the specific nature of the *H* bonding. Thus if one strand has the base sequence: *ATGCCTG*, the other strand must have the complementary bases: *TACGGAC*. This fact, as we shall see later, is of the utmost importance in the translation and replication of the genetic information, but at the moment we are concerned with how this information is carried in the DNA molecule. The Watson and Crick DNA model provides for variability only in the sequence of organic bases along its length, and it is therefore logical to look to the base sequence as the vehicle for transmitting information. This information, it is suggested, is carried in the sequence of the four bases which constitute a four-letter alphabet making up sentences which are of enormous length. Each cell of an organism is equipped with the DNA of that organism, the genotype. The instructions and information necessary for the functioning of the cell are carried in the sequence of the four organic bases within the DNA. The cell must in some way be able to translate that base sequence into the formation of enzymes which will control its metabolism.

2.5. THE GENETIC CODE

The synthesis of the proteins which form enzymes does not take place in the nucleus, where the genetic material is sited, but in the cytoplasm. More specifically the synthesis takes place at minute particles, termed ribosomes, which are carried on the surface of the complex membrane system within the cell, the endoplasmic reticulum. The information carried as a sequence of bases along the DNA molecule must be transferred to the ribosomes, where it is used to control the joining together of amino acids to form proteins. Since there are only four different organic bases present in the DNA molecule, it is clear that one base cannot 'code' for a single amino acid. Neither can a code of base pairs be used, since there are only sixteen combinations involving two bases possible from the four bases, and hence the minimum code will be a triplet of bases for each amino acid. In fact there are sixty-four such base triplets, rather more than three times the number of common amino acids. It will be seen later that almost all the amino acids are coded for by more than one base triplet.

The transfer of the base sequence code from the DNA to the ribosomes involves another nucleic acid, *ribonucleic acid* (RNA). The RNA molecule is very similar to that of DNA, except that the sugar is ribose and the base thymine is replaced by uracil. The RNA molecule is a long, polynucleotide chain but unlike DNA is often found as a single strand. The part played by RNA in the transfer of the genetic information and the synthesis of proteins has been pieced together from the results of several lines of investigation. Radioactive inorganic phosphate injected into an animal is rapidly incorporated into its RNA. If the animal is killed shortly after the injection and liver cells removed and fractionated into the various parts, nucleus, mitochondria and endoplasmic reticulum, it is found that by far the most radioactive RNA is found in the nucleus. If extraction of the liver cells is delayed for some time after the injection the cytoplasm is found to contain a much larger proportion of radioactive RNA. This experiment, and other comparable ones, indicated that at least some of the cell's RNA was manufactured

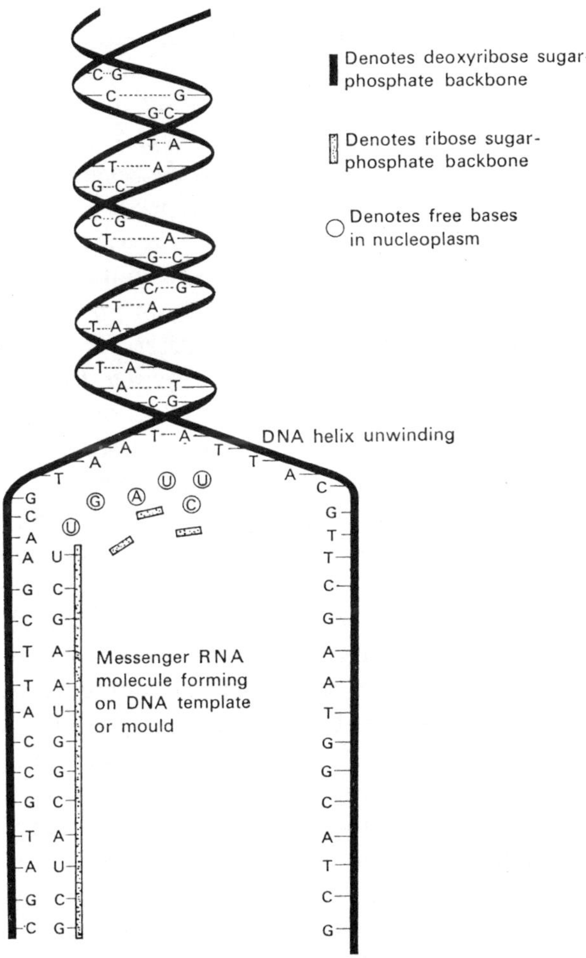

FIG. 19
Formation of messenger RNA on DNA template

in the nucleus after which it moved out into the cytoplasm. Coupled with the similarity in structure between the DNA and RNA molecules, particularly in terms of the presence of a base sequence, this experiment suggested that RNA may be responsible for transferring the base sequence from the nuclear DNA to the ribosomes. This has received confirmation more recently from Astrachon and Volkin who showed that when a bacteriophage injects its DNA into the bacterial cell it causes the cell to produce a new type of RNA. The newly synthesised RNA has a base sequence which mirrors that of the phage DNA, indicating that RNA can be built up at the DNA molecule 'template' and will reflect in its base sequence the base sequence of the DNA. For this process to take place it is clear that the double DNA helix must unwind, and that therefore there must be two DNA strands, with different base sequence, available for copying by the RNA. At the present time it is thought that only one of the DNA chains is used, and always the same one. Thus the first stage in the utilisation of the genetic information takes the form of a 'copying' of the base sequence of the nuclear DNA by a long, single strand of RNA. Since this RNA is responsible for carrying the information to the site of protein synthesis in the cytoplasm it is called *messenger RNA* (mRNA).

The messenger RNA carries its information specifically to the sites of protein synthesis, the ribosomes. The ribosomes themselves consist largely of RNA which is bound to protein. Although it is known that protein synthesis takes place at the ribosomes the exact function of the ribosomal RNA is not known. For protein synthesis to take place the requisite amino acids must be present at the ribosome at the right time. This is the function of a third type of RNA, *transfer RNA* (tRNA). The transfer RNA molecule consists of a relatively short polynucleotide chain which is coiled back upon itself. The manner of coiling means that at one end of the molecule there are three bases which are not paired. Each transfer RNA molecule, with a particular unpaired base triplet, will, in the presence of a specific activating enzyme, combine with a particular amino acid. Thus each amino acid is carried to the ribosome by a transfer RNA molecule which has, at one end, a specific unpaired base triplet.

The messenger RNA becomes attached to the ribosomes. We may think of the messenger RNA as a series of unpaired base triplets, the sequence of which is derived from the DNA. Surrounding the ribosomes are the amino acids, each attached to a transfer RNA molecule with a specific unpaired base triplet. The ribosome moves along the messenger RNA molecule and as each base triplet comes up the transfer RNA molecule with the complementary base triplet locks on to the messenger RNA. Thus a transfer RNA with the base triplet *ACC* will fit into a messenger RNA triplet *UGG*, one with *UCU* will fit into *AGA* etc. This locking together of the complementary triplets on the messenger and transfer RNA ensures that the amino acids are brought to the ribosomes in the sequence dictated by the messenger RNA, and thus by the DNA. While the transfer RNA is linked to the messenger RNA a peptide bond is formed between its amino acid and the preceding one, thereby building up the protein chain. The chain of amino acids is thus passed to the transfer RNA locked to the messenger RNA, and the preceding transfer RNA is released into the cytoplasm to combine with another amino acid. This creates room for a further transfer RNA plus amino acid to lock on to the messenger RNA. In some cases it is found that as soon as one ribosome has moved some distance along the messenger RNA strand another one begins working along it. In this way several ribosomes may utilise the same messenger RNA strand at the same time giving aggregates of ribosomes termed *polysomes*. Clearly this system would give much more rapid production of particular proteins.

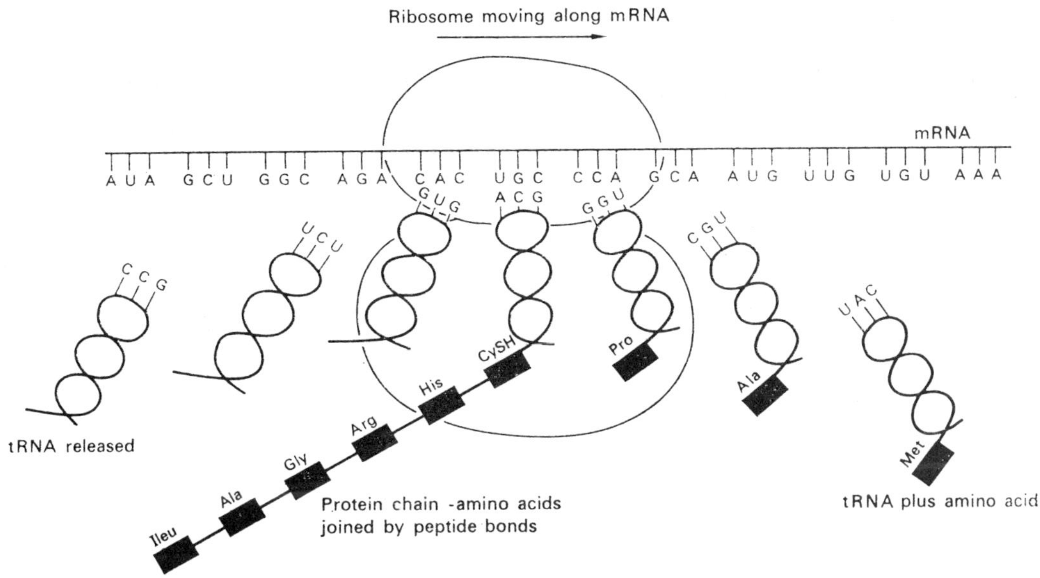

FIG. 20
Diagrammatic representation of protein synthesis at the ribosome

At an earlier stage it was stated that there are far more base triplets than are necessary to code the twenty amino acids. It was found that in most cases an amino acid was coded for by more than one triplet, in fact a number such as leucine and arginine are coded for by no fewer than six base triplets. Others, such as methionine and tryptophan, have only one triplet each. For this reason the code is said to be degenerate. All but three of the base triplets have been shown to code for an amino acid. It now seems almost certain that all these three triplets, *UAA*, *UAG* and *UGA*, are 'nonsense' triplets which have an important function in the genetic code. It is thought that when one of these triplets occurs along a messenger RNA chain it acts as a 'full stop', causing the addition of amino acids to stop and the completed protein chain to be released.

Second letter

First letter	U		C		A		G		Third letter
U	UUU	Phe	UCU	Ser	UAU	Tyr	UGU	CySH	U
	UUC		UCC		UAC		UGC		C
	UUA	Leu	UCA		UAA	Ochre	UGA	?	A
	UUG		UGG		UAG	Amber	UGG	Try	G
C	CUU	Leu	CCU	Pro	CAU	His	CGU	Arg	U
	CUC		CCC		CAC		CGC		C
	CUA		CCA		CAA	GluN	CGA		A
	CUG		CCG		CAG		CGG		G
A	AUU	Ileu	ACU	Thr	AAU	AspN	AGU	Ser	U
	AUC		ACC		AAC		AGC		C
	AUA		ACA		AAA	Lys	AGA	Arg	A
	AUG	Met	ACG		AAG		AGG		G
G	GUU	Val	GCU	Ala	GAU	Asp	GGU	Gly	U
	GUC		GCC		GAC		GGC		C
	GUA		GCA		GAA	Glu	GGA		A
	GUG		GCG		GAG		GGG		G

FIG. 21

The genetic code

From *The Genetic Code III*, F. H. C. Crick. Copyright © (October 1966) by *Scientific American*.

2.6. THE CONCEPT OF A GENE

The material which determines a large number of the phenotypic characters of an organism is contained within its genotype. The genotype can be subdivided into a large number of units, each of which is responsible for a single phenotypic character, or a group of related characters. Various names have been applied to these sub-units at various times, but they are now universally termed *genes*. With our present knowledge of the genetic code it is possible to extend this definition of a gene much further. We have seen that the genotype is contained within the nuclear DNA, and that the information carried as a sequence of organic bases in the DNA is translated to form a specific protein. This protein may be an enzyme, in which case it will be capable of causing certain reactions to take place within the organism which will in turn determine phenotypic characters. Thus we can interpret a gene as a length of DNA with a specific base sequence which will code for a particular enzyme and this enzyme will bring about certain phenotypic characters. In the following section we shall examine some examples of this 'one gene—one enzyme' mechanism, but it must be made clear that there are a number of cases known where more than one gene controls the formation of a single enzyme.

2.7. THE CONNECTION BETWEEN GENES AND ENZYMES

2.7.1. *Alkaptonuria*

About one person in every million suffers from an inherited metabolic disorder called alkaptonuria. The affected individuals produce urine which turns black on standing, due to oxidation of homogentisic acid which is excreted in the urine. Normal individuals convert homogentisic acid to acetoacetic acid, which is again excreted in the urine and which does not turn black on standing. This reaction requires an enzyme which is either absent or inactive in alkaptonurics. The trait is inherited as a single gene, and a change in this gene has produced individuals whose phenotype is affected by the loss of a particular enzyme.

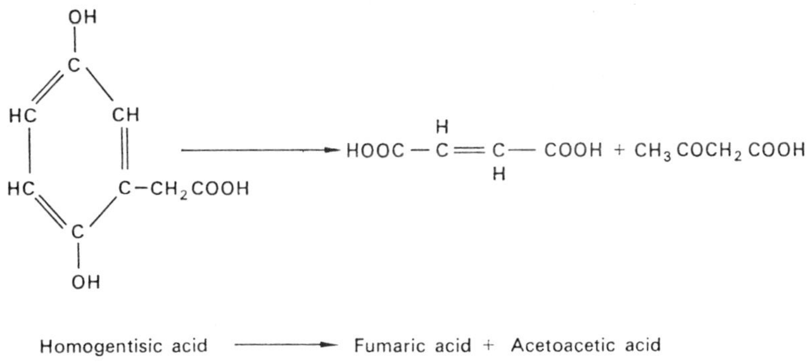

FIG. 22
Metabolism of homogentisic acid

2.7.2. *Phenylketonuria*

In some cases the alteration of a single gene and the subsequent loss, or inactivation, of a single enzyme can cause a variety of phenotypic effects. Phenylketonuria is inherited as a single gene and those affected by the disease excrete phenylpyruvic acid in their urine, are mentally retarded and have much reduced skin pigmentation and quantity of hair. The alteration of a single gene again leads to the functional loss of a single enzyme, in this case the one required to catalyse the conversion of phenylalanine into tyrosine. Instead the phenylalanine is converted to phenylpyruvic acid. The cells of the central nervous system tend to accumulate phenylpyruvic acid, which leads to eventual damage of nerve tissues and subsequent mental retardation. Tyrosine is a normal precursor of the pigment melanin in the body and a block in the production of tyrosine leads to a lack of melanin. Hence phenylketonurics exhibit very fair hair and skin.

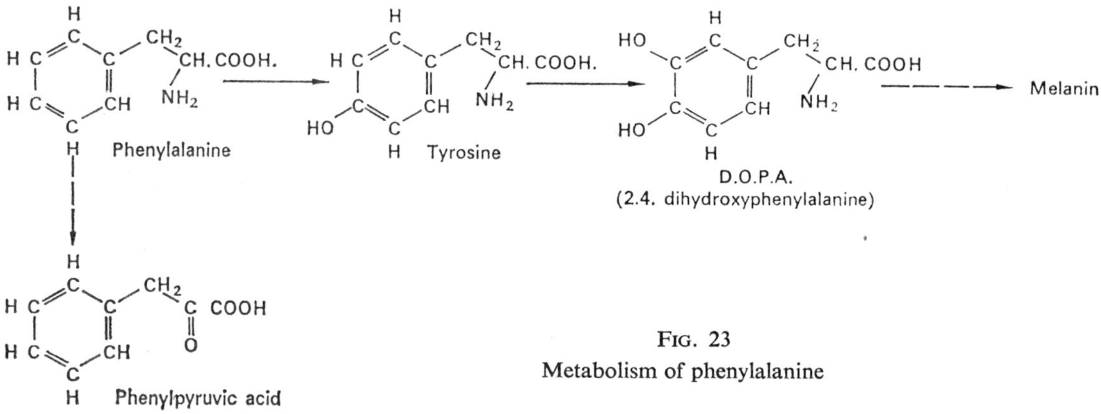

FIG. 23
Metabolism of phenylalanine

2.7.3. *Amino-acid synthesis in Neurospora*

Natural strains of the ascomycete fungus *Neurospora crassa* will grow on a medium consisting of glucose, simple salts and the vitamin biotin. This medium, from which the fungus is able to synthesise all its complex chemical constituents, is termed the 'minimal medium'. In 1941 Beadle and Tatum produced strains of *N. crassa* which differed from the normal type in their inability to synthesise certain amino acids. These strains were produced by the alteration of certain of the organism's genes, a process termed *mutation* which will be discussed in greater detail in Chapter 6. Several *mutant* strains were produced which were unable to synthesise the amino acid arginine. The synthesis of arginine is known to occur in a number of stages, which take place in the sequence:

glutamic acid ⟶ *ornithine* ⟶ *citrulline* ⟶ *arginine*

Thus since the sequence of synthesis is known it is possible to locate at which point the sequence is blocked in each mutant strain. Three different strains were isolated which would not grow on minimal medium but which would grow on minimal medium plus arginine. The three strains were shown to be produced by mutation of three different genes:

Strain 1. grows on minimal medium + ornithine or citrulline or arginine
Strain 2. grows on minimal medium + citrulline or arginine
Strain 3. grows on minimal medium + arginine

Thus it can be seen that in *Strain 1* the sequence of reactions must be blocked between glutamic acid and ornithine, since it is able to convert ornithine through to arginine. Similarly, in *Strain 2* the block must occur between ornithine and citrulline and in *Strain 3* between citrulline and arginine.

Each stage of this sequence of reactions is controlled by a single enzyme and each of the three mutant strains has one of the enzymes either absent or inactivated.

$$\textit{glutamic acid} \xrightarrow{\textit{enzyme 1}} \textit{ornithine} \xrightarrow{\textit{enzyme 2}} \textit{citrulline} \xrightarrow{\textit{enzyme 3}} \textit{arginine.}$$

In *Strain 1* the mutation of the gene concerned has prevented the formation of enzyme 1 and thus blocked the glutamic acid to ornithine reaction. In *Strain 2* mutation of the gene has prevented formation of enzyme 2 and thus blocked the ornithine to citrulline reaction. In *Strain 3* mutation of a third gene has prevented formation of enzyme 3 and thus blocked the citrulline to arginine reaction.

Fusion of cells of two different strains of *Neurospora,* but of the same mating type, results in the formation of a *heterokaryon,* which contains the two genetically different nuclei in the same cytoplasm. When two mutants which are blocked at different points of the arginine synthesis form a heterokaryon, these will grow on minimal medium. A heterokaryon formed from *Strains 2* and *3* would grow on minimal

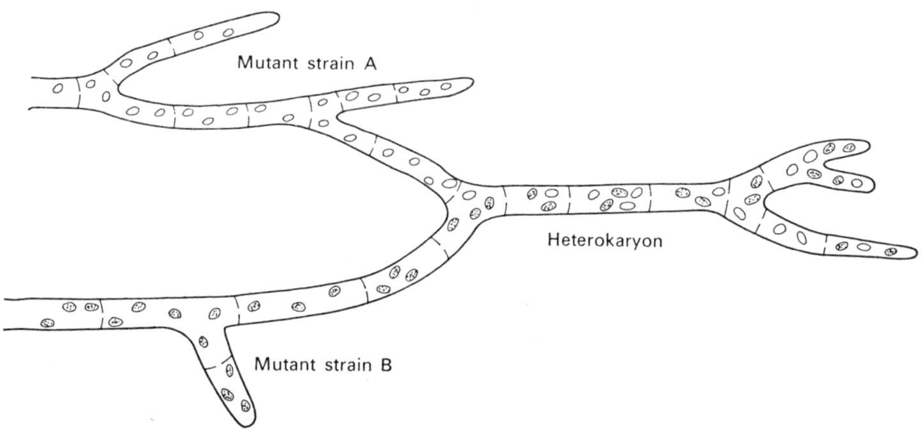

FIG. 24
The formation of a heterokaryon in *Neurospora*

medium since *Strain 2* nuclei will provide information for the production of the enzyme to convert citrulline to arginine and *Strain 3* nuclei will cause the formation of enzymes which will convert ornithine to citrulline. The two genetically different nuclei are said to *complement* each other, and such *complementation* further emphasises the discrete nature of the one gene—one enzyme system.

2.8. THE PROBLEM OF DIFFERENTIATION

We have seen that, with the exception of the gametes, all the cells of an organism contain an identical set of chromosomes, derived from the chromosomes of the zygote. Such is the nature of the process by which this takes place that it seems likely that all the cells contain identical DNA. This DNA would provide a complete set of instructions for the formation of any type of cell within the organism, yet as development proceeds it is clear that individual cells utilise only a small part of the DNA's information. Thus cells will become liver, brain, muscle, connective tissue cells in man, or phloem, parenchyma, xylem, palisade cells in a flowering plant. This process whereby cells utilise only part of the genetic information they contain and become specialised to perform a particular function is termed *differentiation*. The problem of differentiation is why and how only part of the DNA's information is used at a particular time.

Valuable information on how genes may be 'switched on and off' has come from work on the bacterium *Escherichia coli* by Jacob and Monod. They found that the bacterium would grow equally well on a medium containing glucose or lactose, but in the latter case the bacteria produced two enzymes in large quantities, permease and β galactosidase. Permease controls the entry of lactose into the bacterial cell while β galactosidase catalyses the reaction:

$$\textit{lactose} \longrightarrow \textit{glucose} + \textit{galactose}$$

Both these enzymes are present in only minute quantities if the bacterium is grown on glucose, and therefore lactose, in this case, is termed an inducer and the enzymes inducable enzymes. The genes controlling the synthesis of the two enzymes have been

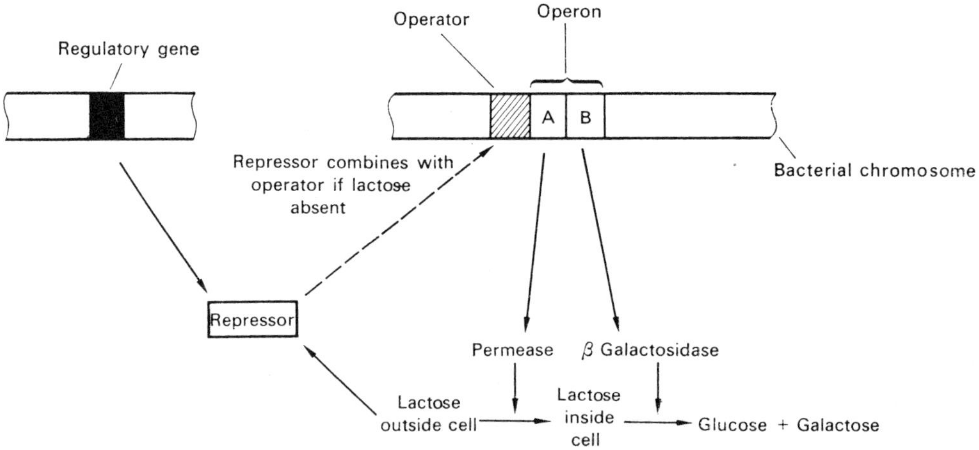

FIG. 25
Induction of enzyme production by lactose in *E. coli*

found to be adjacent on the bacterial chromosome, and since both are required to metabolise galactose they are said to constitute an *operon*. The operon is controlled by an *operator gene* which precedes the DNA coding for the two enzymes. Before the bases of this DNA can be copied on to mRNA the operator must be open. At a different point on the chromosomes there is a *regulatory gene* which produces a repressor enzyme. When the bacterium is grown on a glucose medium the repressor enzyme combines with the operator to prevent the formation of permease and β galactosidase. It is postulated that when the bacterium is grown on lactose, some of this seeps into the cells, combines with the repressor enzyme and thus frees the operator to 'switch on' the enzyme synthesis. This system, and other similar ones which are normally 'switched on', gives a means by which the necessary control over protein production to allow cell differentiation could be exerted.

3
Cell division

3.1. INTRODUCTION

SEXUAL reproduction results in the formation of a zygote, a single cell which contains a nucleus formed by the fusion of a male gamete nucleus with the nucleus of a female gamete. The zygote divides to form two daughter cells, each of which divides forming four cells, then eight and so on until a mass of cells is formed which gradually develops into the young embryo. Growth and development proceed until an adult organism is achieved; the process being basically similar in plants and animals. This means that the multicellular adult, built of a vast variety of different types of tissues, has arisen from one cell.

It has been found that if the nucleus of a zygote is replaced by a nucleus from one of the specialised cells from the adult organism, for example, the nucleus from an intestinal epithelial cell, then the zygote will still develop into the complete organism. This means that the nucleus of a specialised cell still contains all of the genetic information originally found in the nucleus of the zygote. In the previous chapter we found that the molecules of DNA in the chromosomes carry the genetic information, and the amount of DNA present in the somatic cells of an organism is constant. This suggests that prior to the division of a cell the nuclear information is duplicated, one copy being donated to one daughter cell, the other copy passing to the second daughter cell. This type of cell division, where two daughter cells are formed, each containing the same amount of DNA as the parent cell, is called *mitosis*.

3.2. MITOSIS

Mitotic cell division is a continuous process but for convenience it is subdivided into a number of stages.

1. *Interphase*. When the cell is not dividing the nucleus is said to be at interphase. By determining the quantity of DNA in cells at various stages of mitosis it was found that during late interphase the amount of DNA is doubled. Subsequent nuclear division results in a return to the original amount of DNA.

a
Prophase I
Leptotene

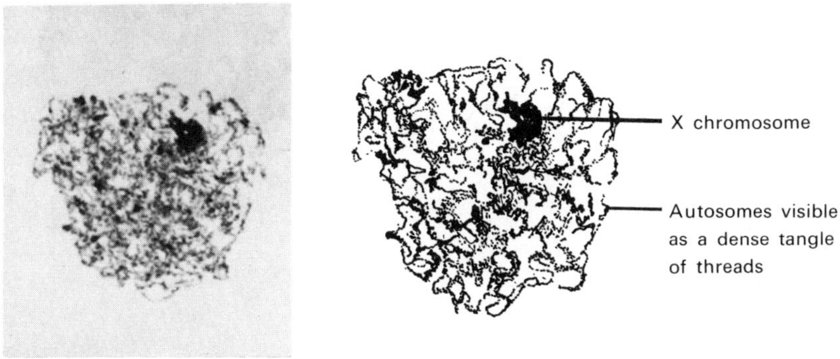

b
Zygotene

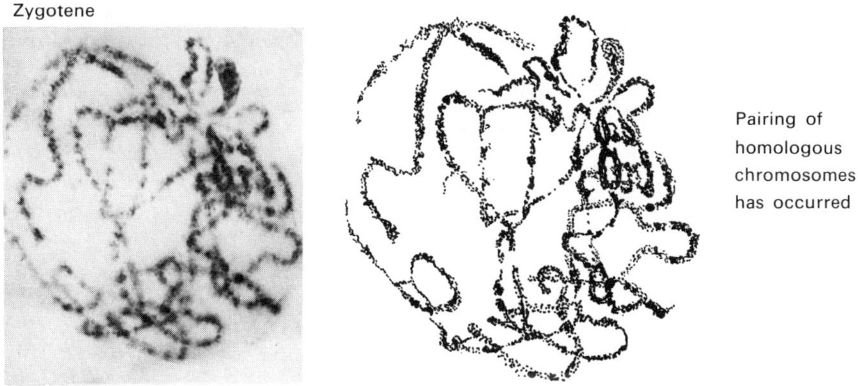

c
Pachytene

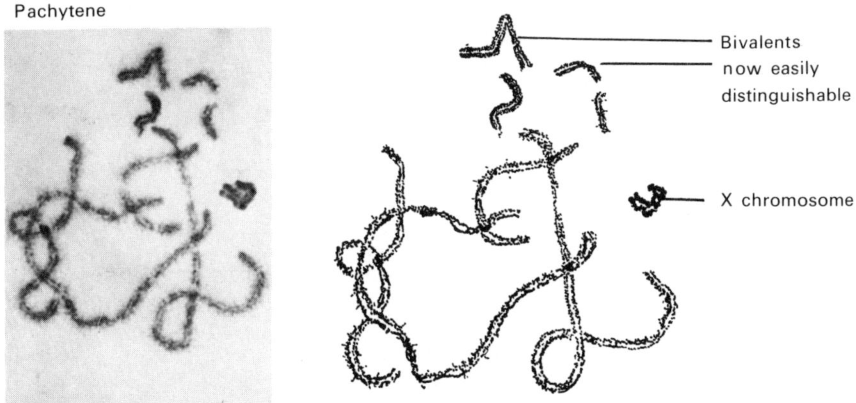

PLATE III a–c
Stages in Meiosis

d
Diplotene

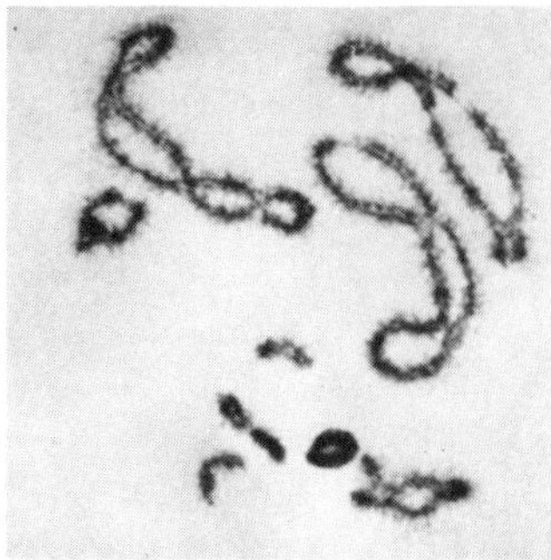

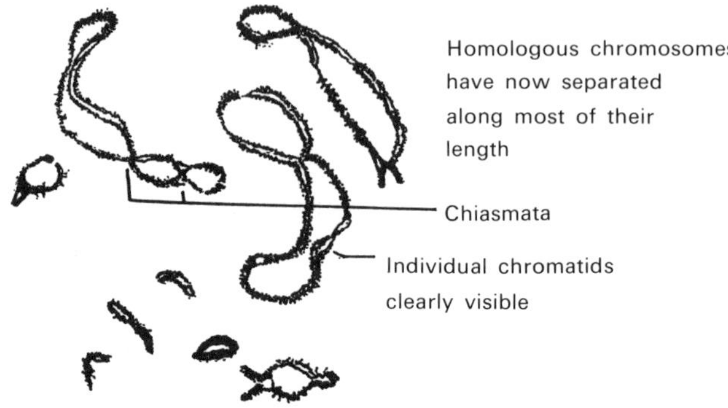

e
Diakinesis/Pro.- metaphase

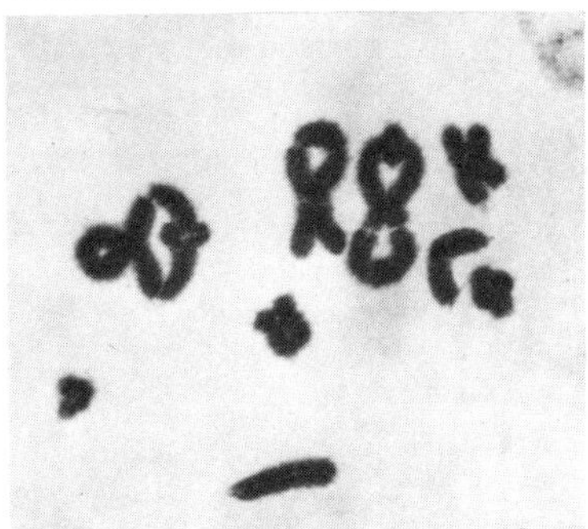

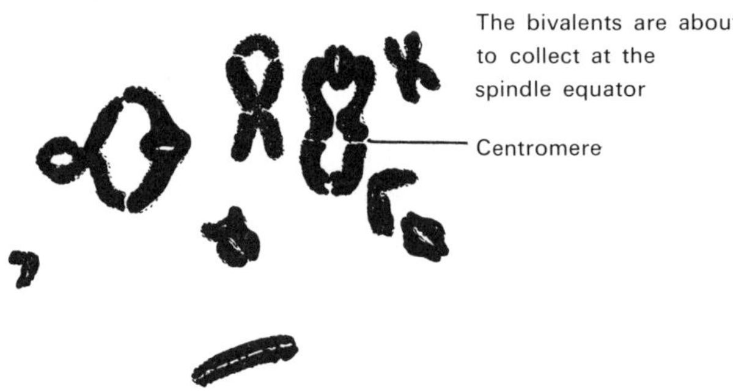

f
Metaphase I

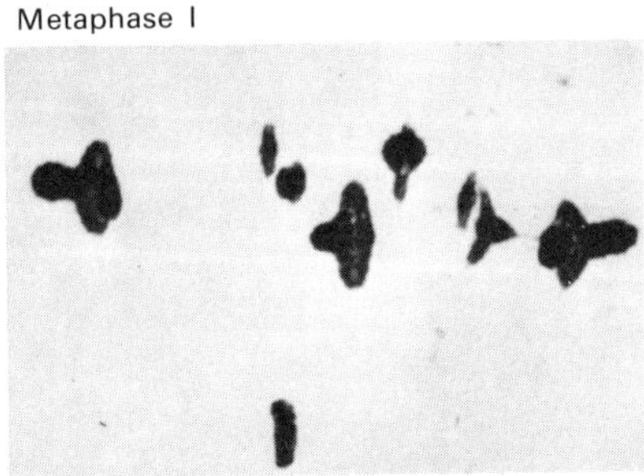

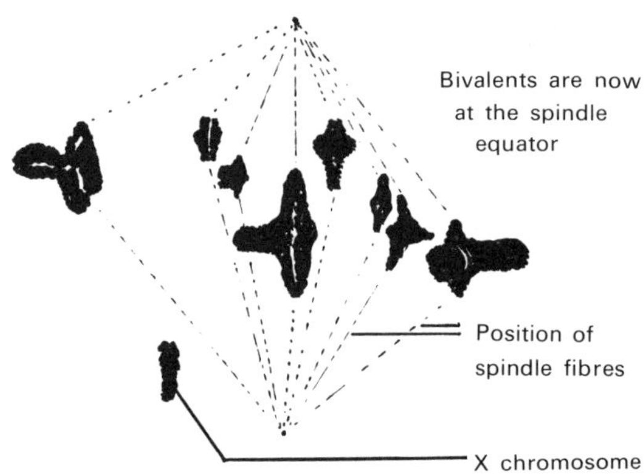

PLATE III d–f
Stages in Meiosis

g
Anaphase I

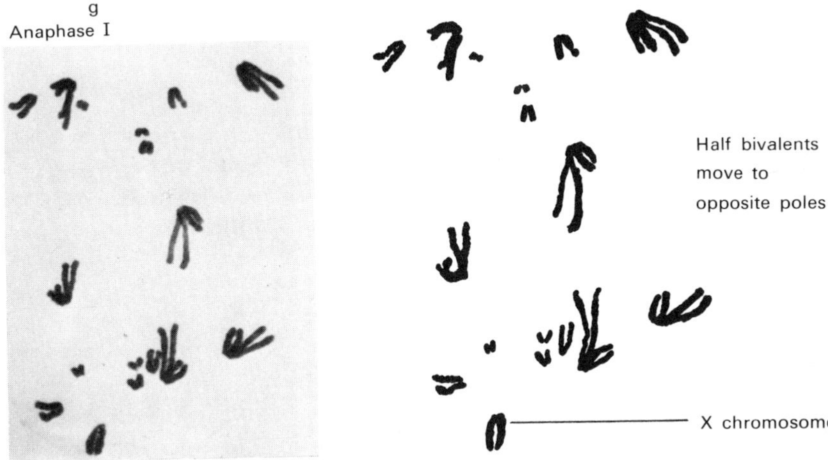

h
Telophase I

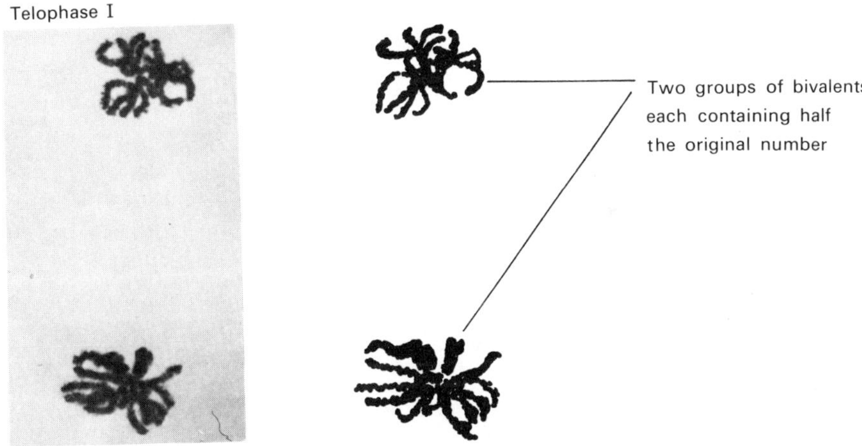

i
Interphase

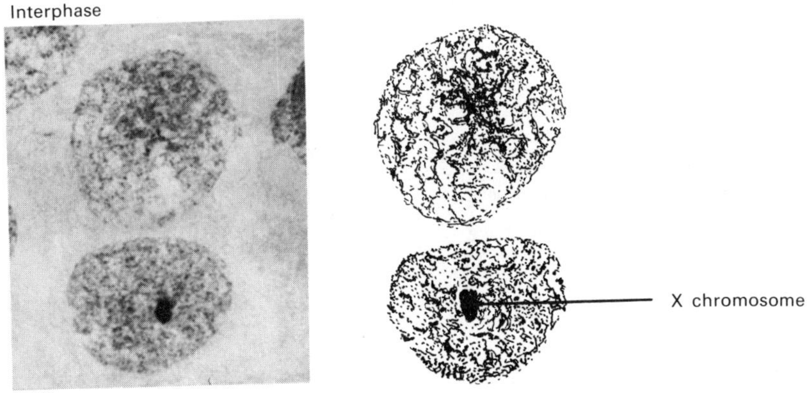

PLATE III g–i
Stages in Meiosis

j
Prophase II

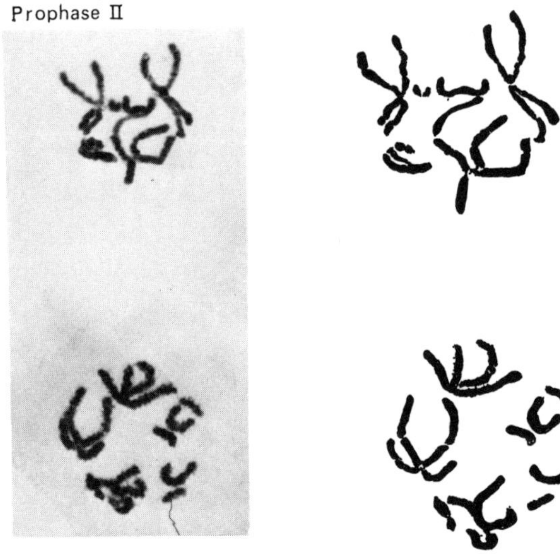

The chromosomes reappear: each consists of a pair of loosly coiled chromatids

k
Metaphase II

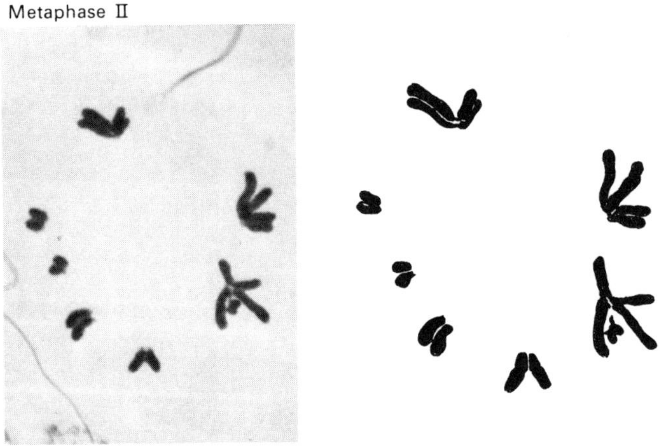

Polar view of chromosomes arranged on the equator

l
Anaphase II

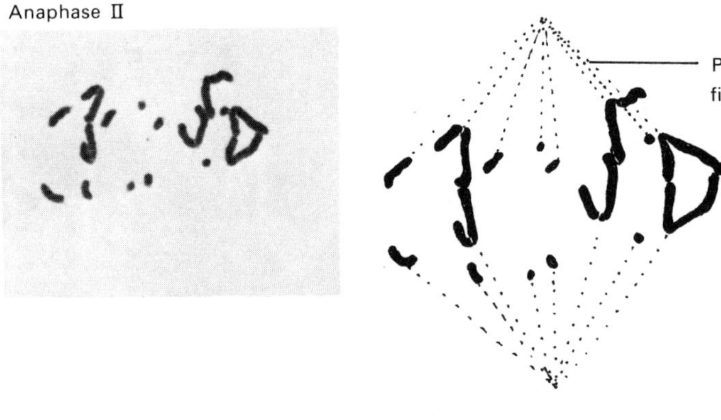

PLATE III j–l
Stages in Meiosis

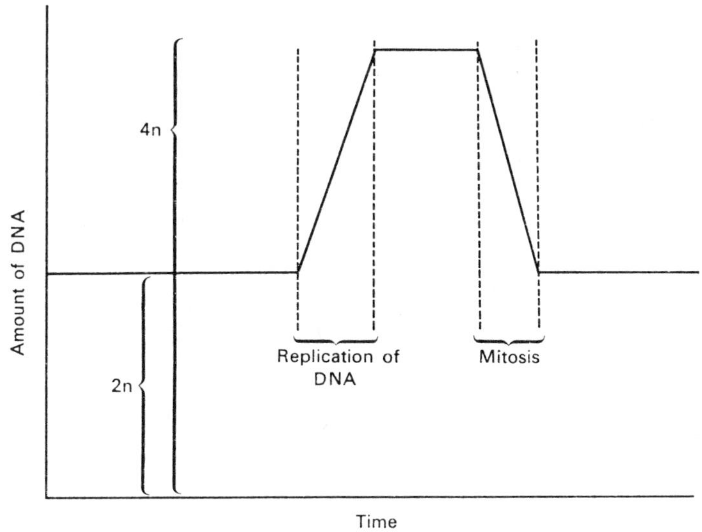

FIG. 26
Duplication of DNA prior to cell division

During interphase the DNA helices and protein links are stretched to their fullest, and the chromosomes are not visible. It is at this time that the RNA, used in protein synthesis, is formed. When the cell is about to divide the formation of RNA is replaced by the synthesis of DNA molecules (Fig. 27). The double DNA helix unwinds at one end. This is due to the breakdown of hydrogen bonds between the organic bases. The separated DNA chains then act as templates for the formation of new chains. The exposed bases pick up their complementary bases from the surroundings. Thus as the bond between an adenine and thymine pair is broken, the adenine links with a free thymine base from the nucleoplasm and at the same time a free adenine molecule links up with the exposed thymine. A new sugar-phosphate backbone is formed, the sugar this time being deoxyribose. In this way as the original double helix of DNA unzips, a new daughter chain is formed along each of the parent chains. Because guanine will only pair with cytosine, and adenine with thymine the new daughter chains will be identical to the chain they have replaced, that is, the correct base sequence will be maintained.

Duplication of all the DNA chains in this manner obviously doubles the amount of DNA in the nucleus. The protein fraction of the chromonemata is also doubled and eventually a second identical 'chromosome' is formed alongside the original. Each of these parallel threads is called a *chromatid* and sister chromatids are connected to each other at one point by a structure called the centromere.

2. *Prophase*. When cell division is about to occur the chromosomes gradually become visible. They shorten and thicken due to each of the chromatids becoming spirally

coiled. The appearance of the chromosomes inside the nucleus is accompanied by the disappearance of the nucleoli. While the chromosomes are shortening changes take place in the cytoplasm. In animal cells a structure called the *centriole* divides and the two resulting centrioles move to opposite ends of the cell. Prophase ends with the breakdown of the nuclear membrane leaving the chromosomes free in the cytoplasm.

3. *Metaphase*. This stage is characterised by the formation of a structure called the *spindle*. This spindle consists of a cone of fibres which radiate from each centriole: the fibres meeting approximately in the centre of the cell give the spindle the appearance of two cones joined base to base. Chemical analysis of the spindle has shown that it consists mainly of protein with some RNA. The protein molecules linked by sulphur—sulphur bonds form the individual fibres of the spindle.

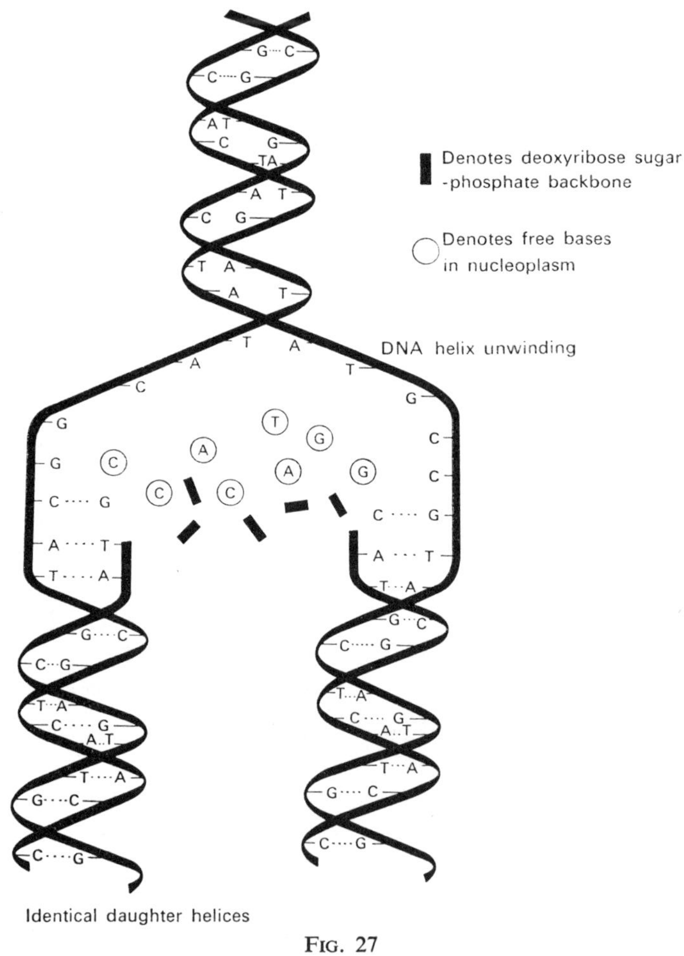

FIG. 27
Duplication of DNA molecule

The widest region of the spindle is called the *equatorial plane*, and the condensed chromosomes come to lie on this equator. Any small chromosomes are usually situated in the centre of the equatorial plane, whereas the larger ones are arranged on the peripheral region. Each chromosome is 'attached' to the spindle by the centromere.

4. *Anaphase.* The sister chromatids now separate at the centromere and begin to peel apart as they move in opposite directions towards the poles of the spindle. Movement of the chromatids up the spindle is probably brought about by a shortening of the spindle fibres attached to the centromeres, followed by an active elongation of the spindle body which completely separates the sister chromatids whose V shape suggests that they are being dragged to the poles of the spindle. The separated chromatids are now called chromosomes.

5. *Telophase.* During this stage the separated sister chromatids form two groups, one at each spindle pole. A nuclear membrane forms about each group, and the chromosomes now uncoil, becoming thinner, longer and less distinct.

Gradually the nucleus returns to the homogenous appearance typical of the interphase condition. In an animal cell, the cytoplasm constricts between the two nuclei and two cells are formed. In plant cells a liquid film forms in the region of the equatorial plane of the spindle. Pectin and calcium pectate are deposited in this cell plate, strengthening it. The cytoplasm on either side secretes a layer of cellulose over the entire surface of the cell plate thus forming a new cell wall.

We have seen the need for mitotic cell division in the development of the adult organism from the original zygote, but its importance does not stop there. Some organisms grow continuously throughout their lives, this type of growth is typical of most plants and here mitosis will be occurring at the growth points in the tips of the stems and roots, the meristems. In many organisms growth in size ceases when the adult stage is reached. This discontinuous growth is typical of many animals. In such an animal mitosis will be replacing damaged or worn out cells, for example mitosis in the Malpighian layer of the skin results in the formation of cells in the epidermis which eventually replace those removed by the constant abrasion of the outer surface. A similar replacement takes place in the cells lining the lumen of the gut, where the passage of roughage in the diet and constant exposure to enzymes calls for the continual renewal of the surface epithelium.

In organisms which employ asexual methods of reproduction, the cells which constitute the offspring will be formed by mitosis of the parent's cells. Thus the interstitial cells of *Hydra* divide by mitosis to form a mass of cells which gradually differentiate to form the daughter bud. In the notches on the serrated leaves of *Bryophyllum* mitosis gives rise to cells which will form the daughter plantlets (see Fig. 1). Because the cells of such offspring are produced by mitosis of the parent cells, the genotypes of daughter and parent will be identical.

3.3. MEIOSIS

The number of chromosomes in the somatic nuclei of an organism is always constant, and this number is the same for all of the members of the same species. Each member will have arisen by mitosis from a zygote. The zygote having been formed by the fusion of a male gamete with a female gamete. If the gametes had the same karyotype as the somatic cells then after fertilisation the zygote and resulting organism would have doubled the normal chromosome number. This is not the case because the karyotype of a given species is always constant. A clue to the explanation can be obtained if the quantity of DNA present in the gametic nuclei is compared with that present in the somatic nuclei (Table 1). There is approximately half the amount present in the somatic cells and this is due to a second type of cell division which reduces the chromosome number to half its original. This reduction division is called *meiosis.*

The karyotype of an organism is made up of two sets of chromosomes, one set from the male gamete, the other from the female gamete. This combination gives the chromosome number typical of the species and is called the *diploid number*. Meiotic cell division reduces the number of chromosomes to half, that is, to the *haploid number*. Examination of the chromosomes of a somatic cell will show that they exist in pairs. Each member of a pair is identical to its partner in length and position of its centromere, giving a characteristic shape. These pairs of similar chromosomes are called *homologous pairs.* One member of each pair is from the female parent, the other is from the male parent.

Meiosis involves two successive divisions of the nucleus, both of which resemble mitosis in general procedure.

3.3.1. *The first meiotic division*

During interphase the chromosomes duplicate themselves by the same process which occurs during mitotic interphase, resulting in the formation of double the amount of DNA.

1. *Prophase.* The start of prophase is again signified by the appearance of the chromosomes due to their shortening and coiling. However, for convenience the first meiotic prophase is further subdivided into a number of phases:

(a) *Leptotene.* During this stage the chromosomes become visible as long threads along the length of which are large numbers of granules called *chromomeres.*

(b) *Zygotene.* Here the chromosomes of homologous pairs come together, a process called *synapsis.* They lie alongside each other in such a way that the centromeres and even individual chromomeres are opposite the corresponding structures in the homologous partner. These pairs of homologous chromosomes are called *bivalents*, and zygotene ends with all of the chromosomes present in the nucleus as bivalents.

(c) *Pachytene.* The chromosomes continue to coil and become very pronounced and readily stained. By now each chromosome is visible as a pair of chromatids joined together at a centromere: thus each bivalent consists of four chromatids, which in some cases can be seen to be coiled around each other.

(d) *Diplotene.* As the shortening proceeds the homologous chromosomes appear to repel each other. Beginning at the centromeres separation occurs along the length of the bivalent, except at certain points. These points are called *chiasmata* (s.g. *chiasma*). The number of chiasmata formed varies with the length of the bivalent. In these regions sections of chromatids have become detached from the parent length and reattached to the non-sister chromatid. Thus there is an exchange of similar sections by non-sister chromatids, a process termed '*crossing over*' (Chapter 5).

(e) *Diakinesis.* This is the final stage of the first meiotic prophase and here shortening of the chromosomes is completed.

2. *Metaphase.* The nuclear membrane now disappears and a spindle is formed in a similar way to the mitotic metaphase. The bivalents now released into the cytoplasm move to the equatorial region of the spindle and become attached to the fibres of the spindle at their centromeres.

3. *Anaphase.* At this point separation of the centromeres, started in Diplotene, is continued as they now move in opposite directions up the fibres of the spindle. Each centromere drags behind it its chromosome which consists of two chromatids. Anaphase ends with homologous chromosomes separated into two groups at opposite poles of the spindle. This may be accompanied by the division of the cell to form two, although this is not always the case.

4. *Telophase.* In some organisms there is no telophase and the two groups of chromosomes now enter the second meiotic division. However, this is not always so and at the end of anaphase a nuclear membrane may form around each group of chromosomes and an interphase condition may follow with the disappearance of the chromosomes. In this case a prophase stage will also occur, as the chromosomes reappear at the start of the second meiotic division.

3.3.2. *The second meiotic division*

1. *Metaphase.* Two new spindles are formed at right angles to the original spindle, one in each daughter cell, or if the parent cell as yet has not divided, at opposite ends of the cell. The chromosomes move to the equatorial region and become attached by the single centromere.

2. *Anaphase.* As in mitotic anaphase the centromeres divide and separate pulling the sister chromatids apart. Movement of the sister chromatids to opposite poles of the spindles now takes place.

3. *Telophase.* The second meiotic telophase results in the formation of four groups of chromatids. Nuclear membranes form about each group and division of the cell into four daughter cells completes meiosis.

The four cells are called a *tetrad* and each contains a nucleus which carries half the original number of chromosomes, the haploid number. In many organisms (the lower plants being an exception) meiosis precedes gametogenesis, and the haploid cells formed by meiotic division would now 'specialise' to form either spermatozoa or ova. Fertilisation would then reconstitute the diploid number. In the lower plants there are often two distinct generations present in a single life history. A gamete-forming, or gametophyte, generation is followed by a spore-bearing, or sporophyte, generation. In this case the cells of the gametophyte are haploid, giving rise by mitosis to haploid gametes, which on fertilisation form a diploid zygote. The zygote divides mitotically to form the sporophyte plant, and then meiosis occurs in the formation of the spores. These haploid spores divide mitotically to form the haploid gametophyte plant.

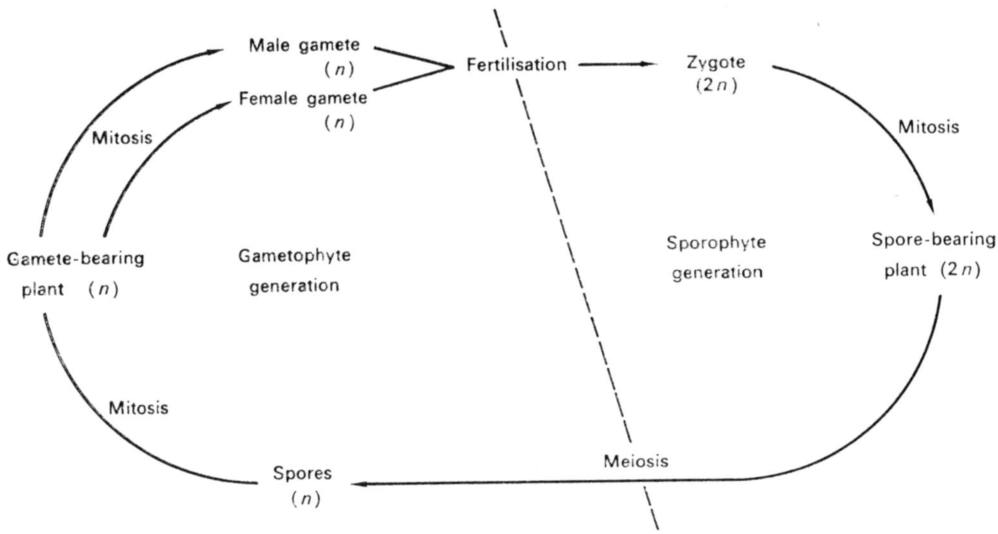

FIG. 28

Alternation of generations in plant life cycles

4
The laws of inheritance

4.1. INTRODUCTION

Drosophila melanogaster is probably one of the most well known of the organisms used by the geneticist. This small fly has a widespread distribution over temperate and tropical areas, and can usually be found about fermenting fruit. The fly complies with the needs of the geneticist in that it is easy to breed, each pair producing hundreds of adult offspring in a relatively short time (ten days at 25°C). This species also has the small karyotype of four pairs of chromosomes (Fig. 29).

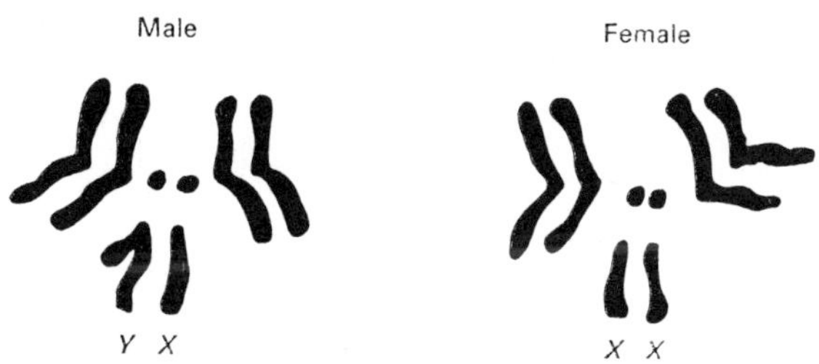

FIG. 29
The chromosomes of *Drosophila melanogaster*

The naturally occurring fly is called the wild type. Over the years variations of the wild type have occurred in laboratory cultures. Many of these variant characteristics are hereditary and can thus be isolated and cultured as true-breeding strains. Today there are large numbers of different strains available. Because each variation is inherited it must have been brought about by some change in the genetic material—that is by mutation (see Chapter 6). Many of the mutations in *Drosophila* involve an alteration in single genes, for example on chromosome 3 there is a gene which, together with others, controls body colour. Mutation of this gene gives rise to a variant with an ebony body instead of the light brown colour present in the wild type. Another gene on chromosome 2 controls, again with others, the shape of the wings; mutation of this gene has produced a strain of flies with very short, crumpled wings, the vestigial strain.

In the diploid somatic cells of the fly, the chromosomes are present in pairs; there will be a pair of chromosomes 1, and so on, giving a total of four pairs. But if there are two number 3 chromosomes and each is identical in the sequence of its genetic material then both must carry a gene for body colour. These genes, representing the same phenotypic character, situated at the same locus on homologous chromosomes are called *alleles* or *allelomorphs.* In a true-breeding strain both alleles must represent the same expression of the character. Thus in a pure strain of ebony body flies both alleles are in the mutant ebony condition: whereas in a pure wild-type strain both alleles will represent the light brown body colour. Where both alleles represent the same variation

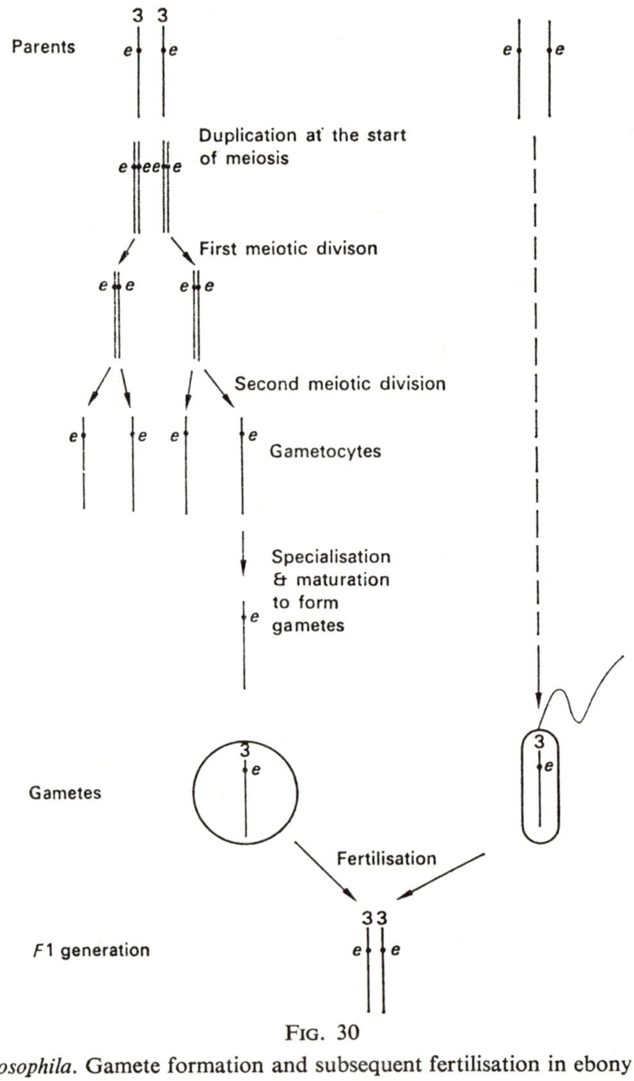

FIG. 30

Drosophila. Gamete formation and subsequent fertilisation in ebony flies of the character, the organism is said to be *homozygous* for that character. True-breeding ebony flies are therefore homozygous for ebony body.

During the formation of the gametes meiosis will form haploid cells by separating homologous chromosomes. Initially duplication of the genetic material occurs so that at the start of meiosis each chromosome consists of a pair of identical chromatids. Homologous chromosomes pair off and the first meiotic division separates these bivalents. The second meiotic division separates the chromatids so that the four haploid cells formed each contain a set of unpaired chromosomes. Applying this to the homozygous ebony flies, at the start of meiosis the four chromatids derived from chromosomes 3 will each carry a gene for ebony body. Completion of meiosis distributes one of the chromatids to each of the four haploid cells formed. These cells will specialise to form the gametes, all of which will therefore contain a gene for ebony body (Fig. 30).

4.2. THE MONOHYBRID CROSS

If an ebony male mates with an ebony female the resulting diploid zygote will contain once again a pair of alleles in the homozygous condition. The offspring will all be ebony (Fig. 30). This first generation of offspring is called the *first filial generation*, abbreviated to *F1*. Now consider a cross between a homozygous wild-type fly and an ebony fly. The wild-type fly will form gametes all of which will be identical in that they will all contain a single allele for wild-type body colour (represented as +). The gametes of the ebony fly will each contain a gene for ebony body (*e*). Fertilisation will reconstitute the diploid condition but this time the alleles representing body colour will not be similar, one allele will be + type, the partner *e* (Fig. 31). Where the pair of alleles

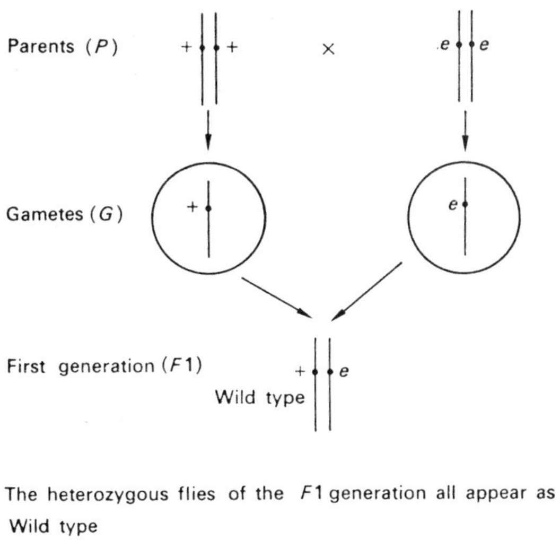

FIG. 31

Drosophila. A cross between a wild-type fly (+ +) and an ebony body fly (*ee*) are dissimilar the organism is said to be *heterozygous* for that condition. In this case the flies of the *F1* generation even though heterozygous for body colour all appear as + type. The effect of the ebony gene is masked by the wild-type gene. It is said that the + gene is *dominant* and the *e* gene is *recessive.*

The heterozygous flies of the *F1* generation will form two types of gamete with regard to body colour. During the first meiotic division the dissimilar alleles will be separated. The second meiotic division will part the chromatids. This means that, of the tetrad formed, two cells will contain the allele for + type body colour, the other two will contain the *e* alleles. The fly will form many hundreds of gametes, and the two types will be present in a 1:1 ratio. Consider a cross between two *F1* heterozygotes. After mating the sperm will come into contact with the ova: '*e*' sperm will stand an equal chance of fertilising an '*e*' egg to form a homozygote, or an '+' egg to form a heterozygote. These fertilisations will occur in approximately equal numbers. The fusion of '+' sperms with both '+' and '*e*' ova will also occur in approximately equal numbers. This will give a final ratio of one ++ to two +*e* to one *ee*. The second generation (*F2*) will have three wild-type flies for every one ebony fly (Fig. 32).

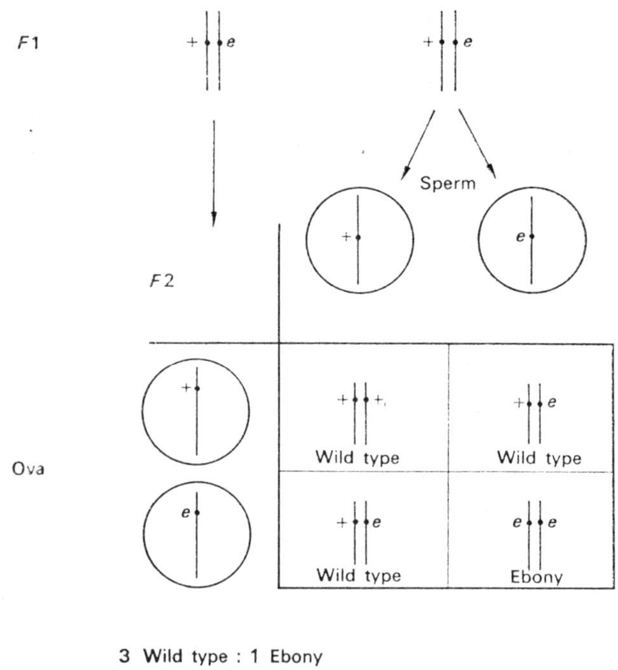

FIG. 32

Drosophila. A cross between two flies both heterozygous for ebony body (+*e*)

In the above diagrams the alleles have been shown situated on their chromosomes, this has been done to simplify the explanation, and is not necessary. Normally the genes alone are shown. When a cross is represented in the tabular form, the recessive gene is indicated by its small initial therefore ebony will be *e*. The dominant wild-type gene is indicated with a +. In cases where there is no recognised wild form the dominant gene is abbreviated to its capital initial therefore tallness in peas will be *T*, and the recessive allele is indicated by the small initial of the dominant, that is *t* will represent the gene for short plants.

The vestigial wing strain (*vg*) in *Drosophila* is also recessive to the wild form. Figure 33 shows the cross between a homozygous wild-type fly and a homozygous vestigial fly. Note the formation of an *F1* generation which all appear wild type even though their genotype is heterozygous. By inbreeding an *F1* male with its sister the

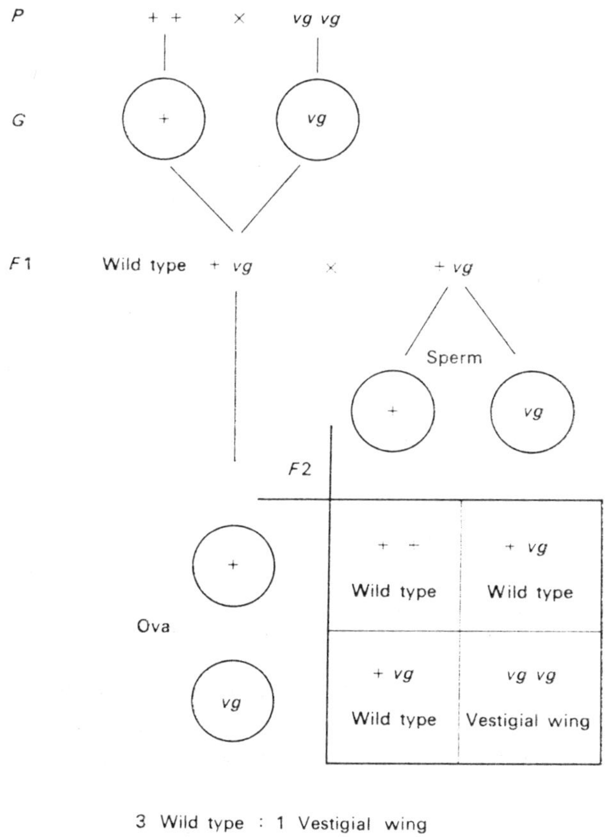

FIG. 33

Drosophila. A cross between a wild-type fly (+ +) and a vestigial wing fly (*vg vg*)

F2 will again show the phenotypic ratio of 3:1. Two-thirds of the wild-type flies will, however, have a heterozygous genotype.

In tomatoes the gene for round fruit, *R*, is dominant to the allele, *r* for elongate fruit. Figure 34 shows a cross between plants true breeding for round fruit and plants

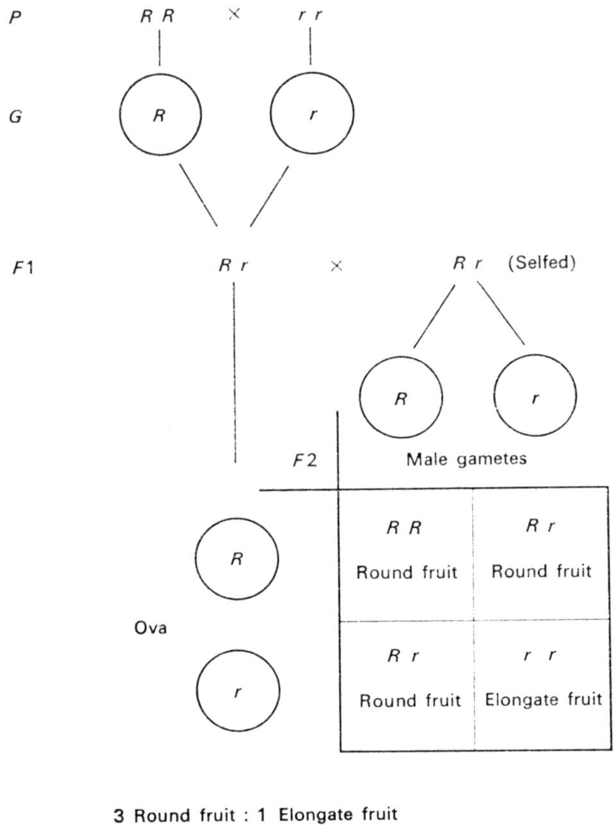

FIG. 34

Tomato. A cross between a plant homozygous for round fruit (*RR*) and a plant homozygous for elongate fruit (*rr*)

true breeding for elongate fruits, with subsequent selfing of the *F1* generation. Again the 1:2:1 genotype ratio, and 3:1 phenotype ratio appear; this is characteristic of a cross involving alleles at one locus, called the *monohybrid cross.*

4.3. THE WORK OF MENDEL

The 3:1 ratio in crosses involving one pair of characters was first noted by Gregor Johann Mendel who was Abbot at a monastery in Brünn (now Brno in Czechoslovakia).

He lived from 1822 until 1884, and trained at Vienna University, studying Physics, Mathematics and Natural Sciences. Prior to being made Abbot, Mendel was a teacher mainly of physics at the monastery, and during this time he experimented with various plants, growing different varieties in the cloister gardens. Mendel partly disagreed with aspects of Darwin's work which was at that time coming into prominence and he began experiments on the edible pea, *Pisum sativum*. The experiments continued for eight years, and in 1866 he published a paper on 'Experiments in Plant Hybridisation'. His work went unnoticed until sixteen years after his death when it was rediscovered, confirmed and accepted.

The garden pea was a fortunate choice by Mendel. It had the advantages that it was easy to cultivate and that its floral structure allowed easy control of pollination (Fig. 35). He found a number of varieties which differed in easily recognised true

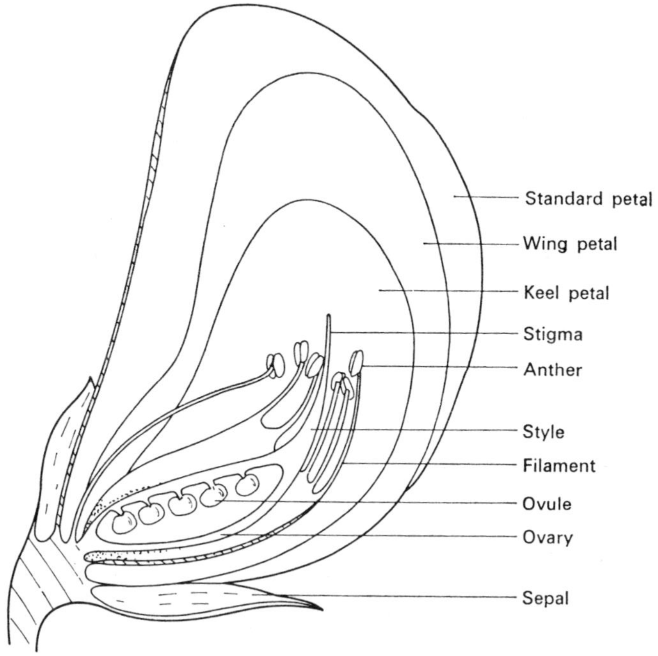

FIG. 35
Half section of the flower of the garden pea, *Pisum sativum*

breeding traits. The pea is normally self-pollinating, pollination occurring before the flower opens, yet the different varieties were inter-fertile and yielded fertile progeny.

Seven of the pairs of contrasting characters that Mendel studied were:

(i) the shape of the dry seed—round or wrinkled;
(ii) the colour of the dry seed cotyledons—yellow or green;

(iii) the shape of the pods—smooth or constricted (between the seeds);
(iv) the colour of the unripe pods—green or yellow;
(v) the position of the flower—axial or terminal;
(vi) the length of the stems—tall or short, and
(vii) the flower colour—purple or white.

Mendel's technique was to select the plants he intended to cross, and remove the undehisced anthers from the immature flower of one variety. He would then transfer pollen from the second variety and artificially pollinate the first. The seeds were collected and sown providing the *F1* generation of plants. The appearance of the characters under consideration was noted together with the numbers of the different types. By allowing these plants to self-pollinate naturally, he obtained the seeds of the F2 generation. Again the appearance and numbers of the *F2* plants were noted.

One of the traits studied by Mendel was that of height. He found that both the tall and short varieties always bred true. When he crossed the tall plants with the short plants the *F1* generation were all tall plants. The *F1* were selfed. The short plants reappeared in the *F2* generation in the ratio of 3 tall:1 short. In the *F3* generation, produced by selfing the *F2* plants, he found that all the short plants bred true; the tall plants, however, were of two kinds. One third of the tall plants bred true, but the other two thirds produced both tall and short plants in the 3:1 ratio. From these results Mendel deduced that the *F2* generation contained three types of plant; true breeding tall, hybrid tall and true breeding short, and these occurred in the ratio of 1:2:1 (Fig. 36).

Mendel explained the results by saying that the gametes contained character-producing factors. The offspring receives two sets of factors during fertilisation; one

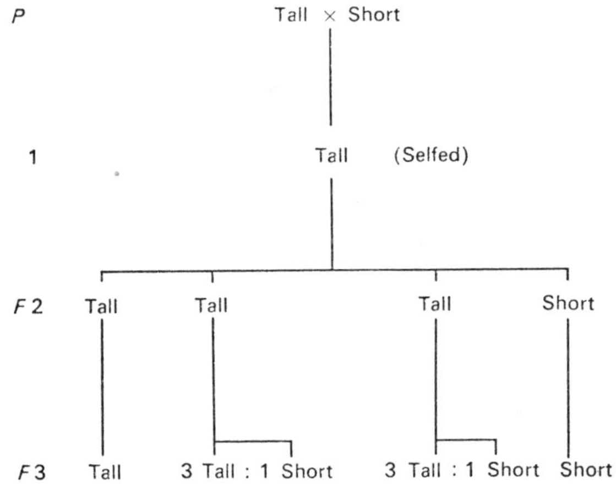

FIG. 36
Mendel's cross between the tall and short varieties of peas

set from the male gamete and the other set from the female gamete. In the above experiment the gametes of the tall plant carry the factor for tallness and the factor for shortness is present in the gametes of the short plant. The zygote, and therefore the *F1* hybrid generation, will carry both factors, however only one, the dominant factor, shows itself.

From his experiments Mendel formulated his law '*The segregation of germinal units*', in which he concluded that during the formation of the gametes opposite factors (pairs of alleles) are segregated, and only one will be represented in a single gamete. Thus in a hybrid plant or animal (hybrid meaning heterozygous), two types of gamete will be formed in equal numbers.

Considering the lack of any relevant cytological knowledge at the time, it is amazing how accurate Mendel's conclusions were. Mendel's 'factors' in the cells are now known to be genes, and the 'segregation of the opposed factors' is a very accurate summary of the meiotic separation of the homologous chromosomes bearing their alleles, forming the 'single set of factors' in the gametes, i.e. the haploid condition. Fertilisation reconstitutes the diploid condition, that is, the 'two sets of factors in the zygote'.

4.4. SEGREGATING ALLELES IN GAMETOGENESIS

The segregation of pairs of alleles to form two types of gamete can be observed in the waxy character in Maize. This character gives the kernels a waxy appearance but is more easily identified by staining with iodine. With such treatment the waxy kernels give a red colour whereas the non-waxy plants give the typical blue colour. The variation is due to a difference in the starch formed in the two types of plant.

If waxy and non-waxy plants are crossed the *F1* are all non-waxy. Selfing produces an *F2* generation in the ratio of 3 non-waxy:1 waxy. This shows that this characteristic is controlled by one gene and waxy is recessive.

The peculiar starch formation is not limited to the kernels but will also occur in the pollen grains according to their genotype. Thus in a homozygous waxy plant, meiosis in the anthers will give rise to haploid gametes each carrying a single allele for waxy character. All of the pollen grains will therefore stain red on treatment with diluted iodine. In a hybrid plant (+ *wx*) segregation of the alleles during meiosis will form two types of gamete, half will contain the allele + the other half *wx*. Thus pollen from a heterozygous plant on staining with iodine should show a 1:1 ratio of blue starch (+) to red starch (*wx*). Plate IV is a photograph of such a preparation. We suggest you count up the numbers of each type and verify the 1:1 ratio using the chi-squared test.

If such a ratio exists it means that during gamete formation segregation does take place and in a heterozygote the two types of gamete are formed in equal numbers.

4.5. THE DIHYBRID CROSS

Besides experimenting with the inheritance of single characters Mendel also worked on pairs of differences and their behaviour in the same cross. For example, he crossed plants grown from round yellow seeds with plants from wrinkled green seeds. The resultant seeds were all round and yellow. This means that the *F1* generation were heterozygous for both pairs of alleles and the allele for round seed is dominant over the allele for wrinkled seed. The allele for yellow cotyledons is also dominant over the allele for green cotyledons.

Mendel grew the *F1* plants and selfed their flowers. The resultant *F2* seeds were of four types in a definite ratio of:

9 round yellow
3 round green
3 wrinkled yellow
1 wrinkled green

In all of his experiments with two pairs of characters Mendel constantly obtained the 9:3:3:1 ratio. From these results he postulated his second law called '*The principle of recombination*' which stated that 'all the various pairs of factors present in an organism segregate independently of one another, and may recombine with either factor of another pair of contrasted characters'.

The explanation of this law becomes clearer if we consider the first meiotic spindle of the *F1* plant during the formation of its gametes. Separation of the chromosomes bearing the allele *R* from the homologue with the allele *r* will occur. The alleles *Y* and *y* will also be segregated as their chromosomes are pulled apart by the spindle. The allele *R* is pulled to the same pole as *Y*, their partners *r* and *y* will be drawn to the opposite poles forming two types of gamete *RY* and *ry*. These are similar to the parental gametes and are, therefore, termed *parental types.* On the other hand it is just as likely that the allele *R* will be drawn to the same pole as the allele *y*. and thus *r* with *Y*—forming another two types of gamete *rY* and *Ry*. These are formed by the recombination or reshuffling of the parent alleles and are called the *recombinant types.*

The four types of gamete will be formed in equal numbers. On selfing, each of the four types of male gamete stands a similar chance to the others of fusing with any of the four types of female gamete (see Figure 37).

If the phenotypes in the chequer-board are totalled these theoretical combinations reflect the 9:3:3:1 ratio Mendel observed in his experiments.

Mendel's second law, although applicable to the above cross and to many other cases, is not universal. Independent segregation and recombination of pairs of factors

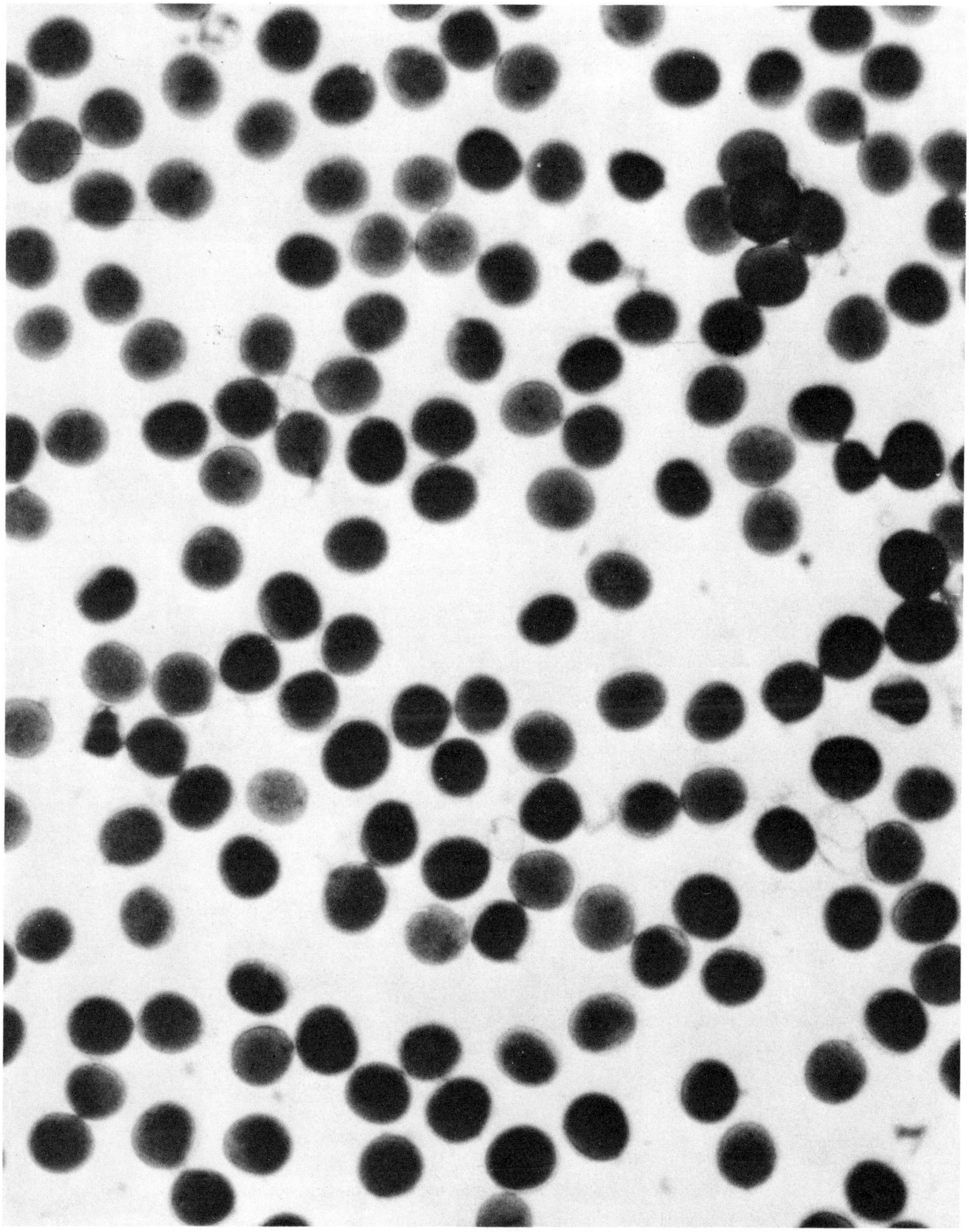

PLATE IV

Pollen from a maize plant heterozygous for the non-waxy/waxy alleles (+ *wx*). After staining with dilute iodine solution pollen with the non-waxy allele stains dark blue, pollen with the waxy allele stains red. In the plate pollen grains stained blue appear darker.

a

b

c

d

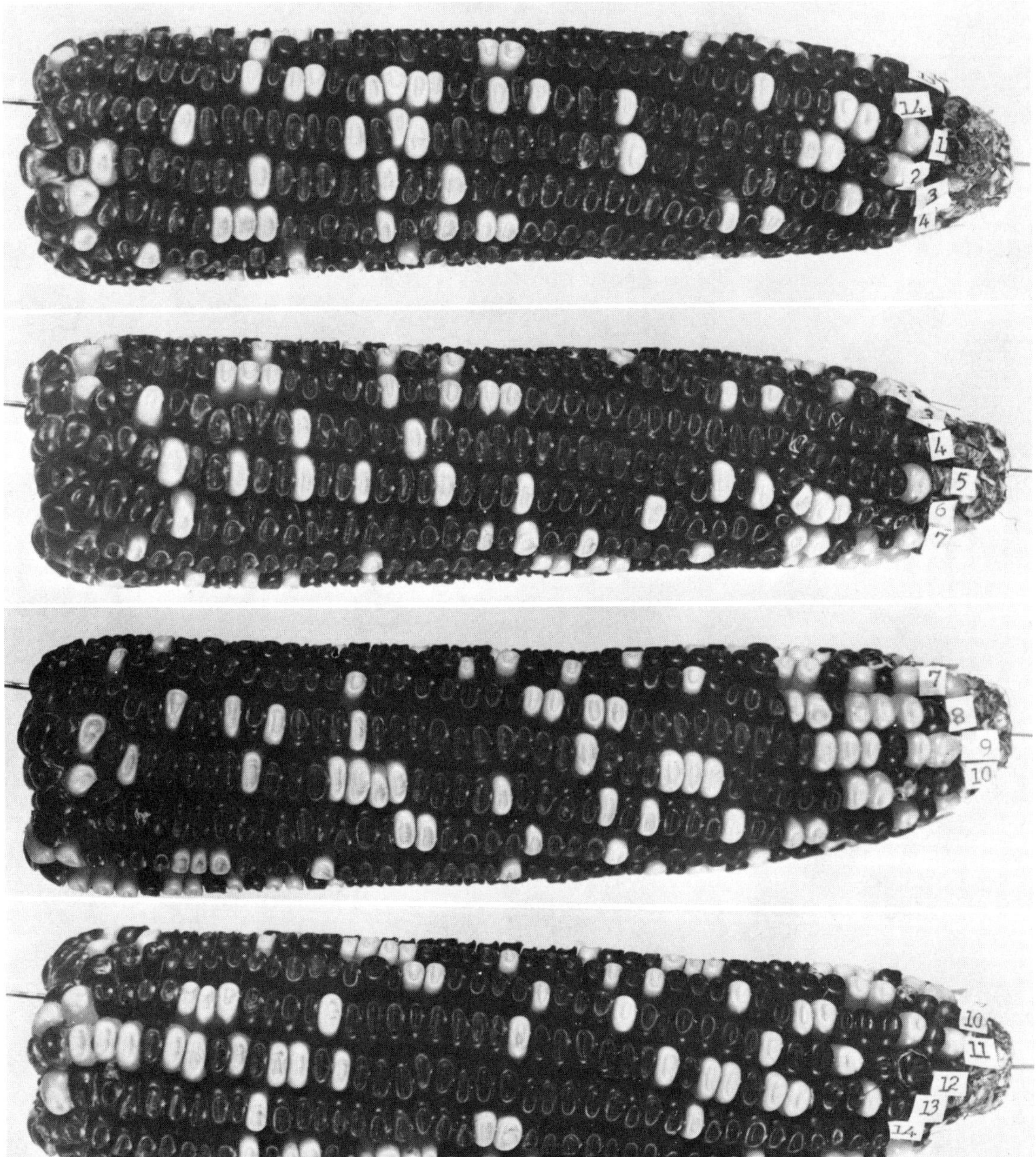

PLATE V a–d

An *F2* hybrid maize cob showing segregation of purple and yellow alleles.

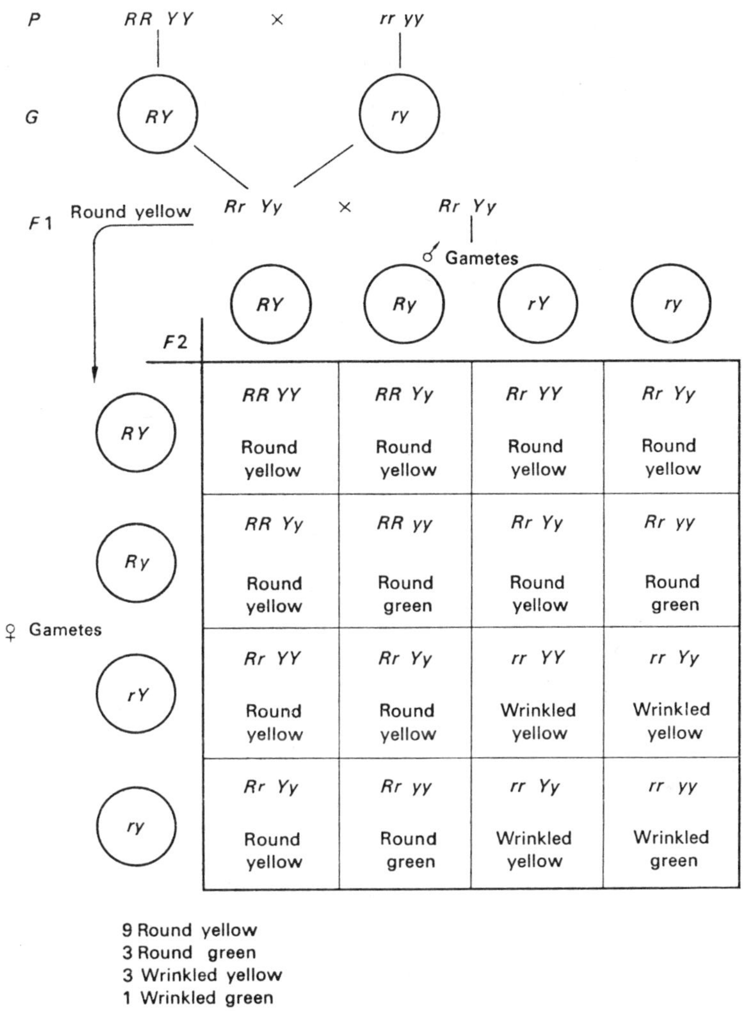

FIG. 37

Pea. A cross between round yellow pea (*RR YY*) and wrinkled green pea (*rr yy*)

will only occur if the genes under consideration are on separate chromosome pairs. If both pairs of alleles are on the same homologous chromosomes then independent segregation will not take place in the majority of cases. The genes will behave as though they were linked together. This phenomenon called 'linkage' is dealt with in Chapter 5.

4.6. FURTHER EXAMPLES OF DIHYBRID CROSSES

Consider the case of *Drosophila* mentioned at the beginning of this chapter. Wild-type wing and body colour are both dominant to the vestigial wing and ebony body mutants. A dihybrid cross to the *F2* generation would produce the results shown in Fig. 38.

The same *F1* and *F2* are produced if the pairs of recessive alleles are in separate parents. Consider the cross between short-eared mice (+ + *se se*) and mice carrying

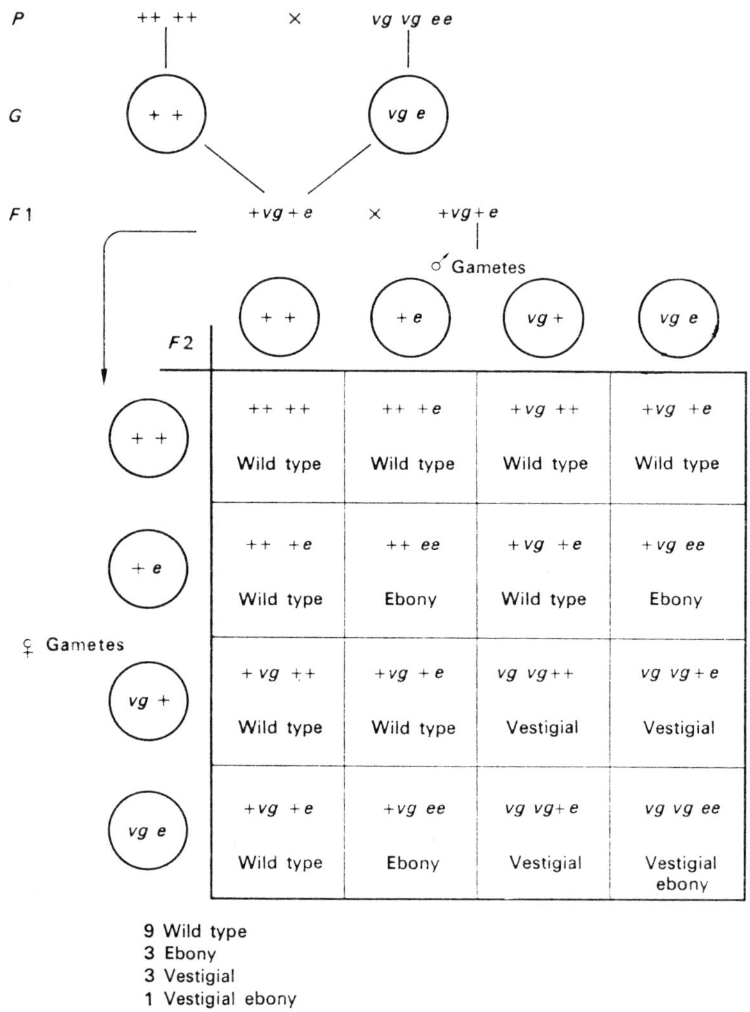

FIG. 38

Drosophila. A cross between a wild-type fly (+ + + +) and a vestigial ebony fly (*vg vg ee*) the gene for pink eye dilution ($p\,p\ +\,+$) which also have a sandy coat. The results are shown in Fig. 39.

4.7. THE TESTCROSS

In hybrid organisms, where one dominant allele masks the effect of the recessive allele, the appearance of the heterozygote will be identical to that of the homozygous dominant organism. Consider again the earlier example in tomato plants. The plants of

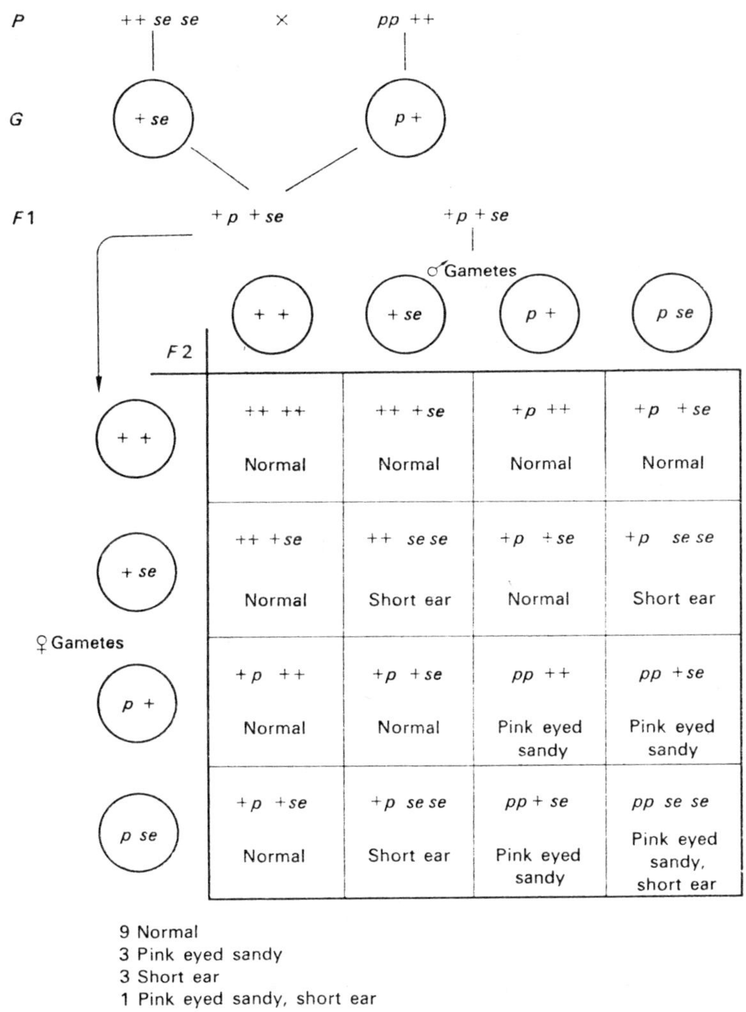

FIG. 39

Mice. The results of crossing a short ear mouse (+ + *se se*) with a pink-eyed sandy mouse (*pp rr*) genotype *RR* produce round fruit, but so do plants with a genotype *Rr*. The question arises 'if presented with a set of plants containing both genotypes, how can we distinguish between them if their appearance is identical?' The simplest way is to cross the plants with the homozygous recessive strain. Crossing the homozygous dominant plant, *RR*, with the recessive strain, *rr*, will produce heterozygous offspring, *Rr*, which all appear identical, for instance see Fig. 40.

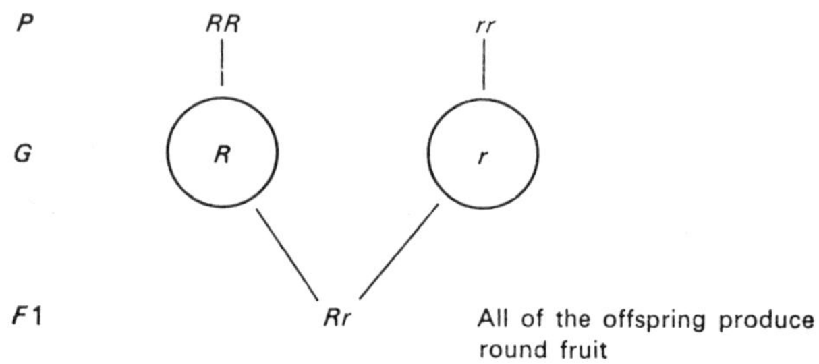

FIG. 40
Testcross of a tomato plant producing round fruit (*RR*)

However, on crossing the heterozygous plant with the homozygous recessive both types of phenotype will appear in the offspring in a 1:1 ratio (Fig. 41).

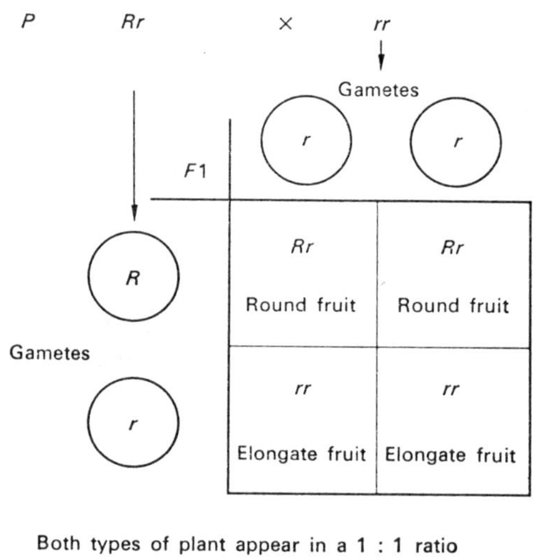

FIG. 41
Testcross of a tomato plant producing round fruit (*Rr*)

This is called a testcross and can also be applied to the offspring of a dihybrid cross. The organism to be tested is this time mated with the double recessive strain.

If the test organism is homozygous for the dominant genes then only one type of offspring will appear in the *F1* (Fig. 42).

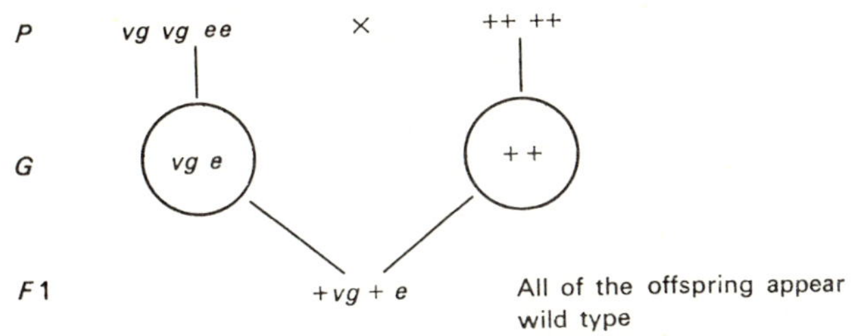

FIG. 42
Drosophila. Testcross of a fly with a wild-type phenotype (++ ++)

The test organism may, however, be heterozygous, in which case on crossing it with the double recessive, four types of offspring will appear in a 1:1:1:1 ratio (Fig. 43).

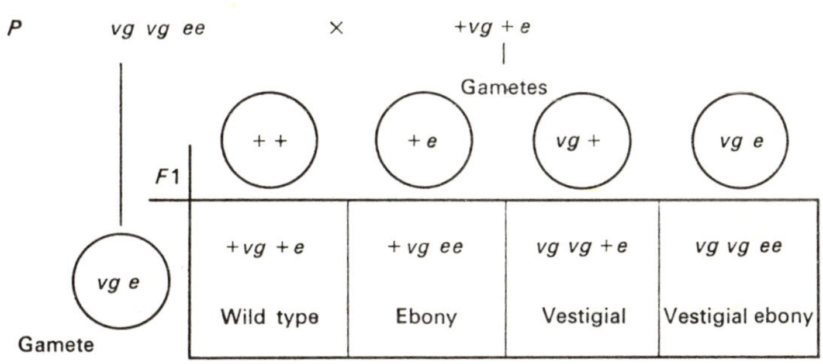

1 Wild type
1 Ebony
1 Vestigial
1 Vestigial ebony

FIG. 43
Drosophila. Testcross of the double heterozygote (+ *vg* + *e*)

4.8. INCOMPLETE DOMINANCE

Complete domination by one allele over its partner is not always the case. In some instances the heterozygote is clearly distinguishable from both of the homozygotes. The Andalusian fowl provides an example. If a black fowl is crossed with a white bird the *F1* are all blue. The feathers of the *F1* bird have a fine pattern of white and black giving a blue effect overall. This means that dominance is absent and the offspring are intermediate between the two parental types. Inbreeding of the blue *F1* generation produces three types of fowl in the *F2*: 1 black:2 blue:1 white—see Fig. 44.

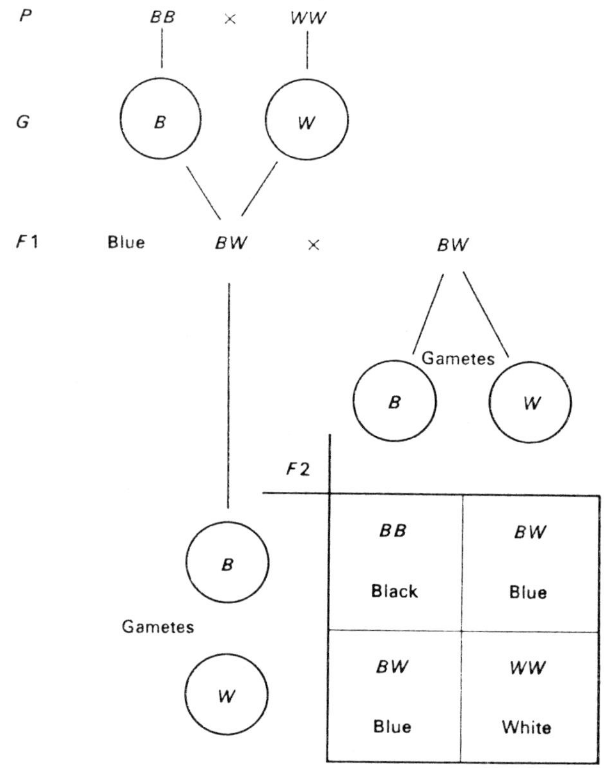

FIG. 44
Incomplete dominance in Andalusian fowl. Results of crossing a black bird (*BB*) with a white bird (*WW*)

Another example of incomplete dominance is found in the Snapdragon (*Antirrhinum*). If a red-flowered plant is crossed with a white-flowered plant the resultant heterozygotes are all pink-flowered. On selfing the pink *F1* generation, the three phenotypes reappear in the *F2* generation in the modified 3:1 ratio of 1 red:2 pink:1 white (Fig. 45).

Once again the heterozygote is clearly distinguishable from both the parents, that is neither allele is dominant.

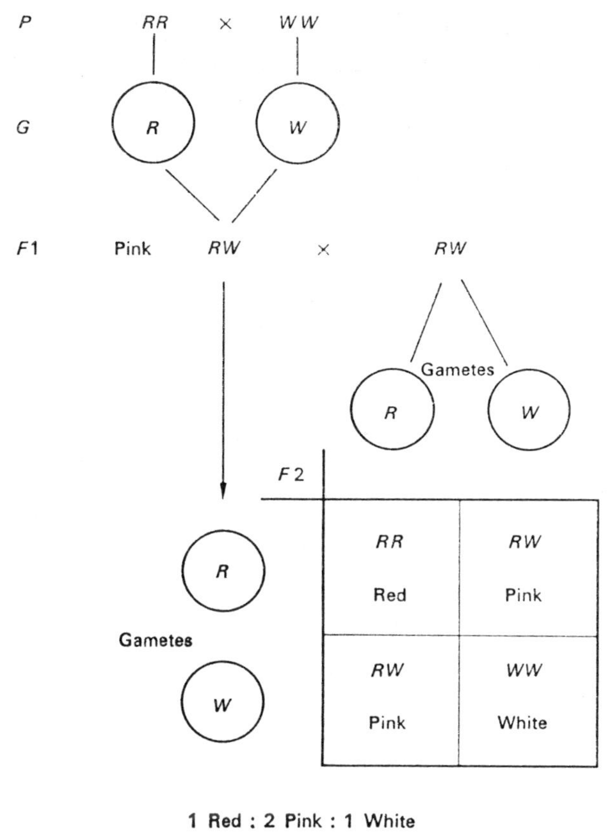

FIG. 45

Incomplete dominance in *Antirrhinum.* Results of crossing a red-flowered plant (*RR*) with a white-flowered plant (*WW*)

4.9. GENE INTERACTION

So far in this chapter we have considered a gene as a portion of the chromosome controlling one particular character. This however is not always the case: many instances are known where a gene may influence several separate characters, a phenomenon known as *pleiotropism* (see later in this chapter). A second implication in the above statement is that a given character is only controlled by alleles at one particular locus. This again is not so.

In poultry the shape of the comb is modified by at least two genes at different loci—*P* and *R*. The familiar high serrated comb, called the 'single' comb, is produced by the homozygous double recessive *pp.rr*. The dominant allele *P* gives a 'pea' comb which is shorter than the 'single' and has rounded lumpy projections along the middle line. The second dominant gene R gives a 'rose' comb consisting of a triangular mass of small spikes; the apex of the triangle points backwards and is free from the head. Where both dominant genes are present in the genotype a 'Walnut' comb is formed. This has no distinct spikes as in the 'pea' or ridges like the 'rose' instead it consists of a wrinkled sometimes hairy lobe of tissue (Fig. 46).

FIG. 46
Comb shape in poultry

A cross between a homozygous 'rose' and a homozygous 'pea' comb bird once again produces Mendel's dihybrid ratios—shown in Fig. 47.

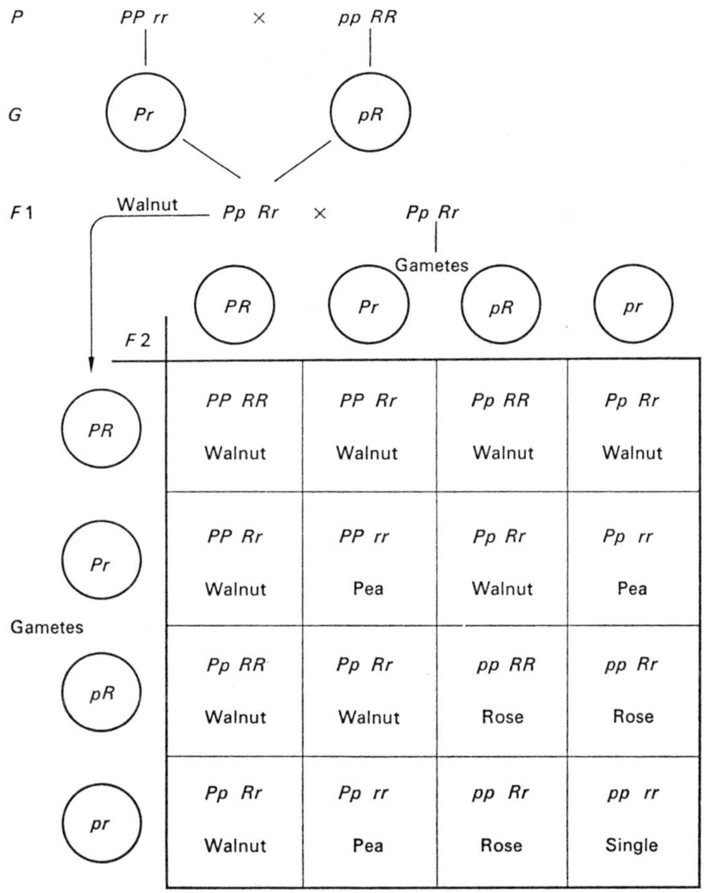

9 Walnut
3 Pea
3 Rose
1 Single

FIG. 47

Gene interaction in poultry. Results of crossing a pea comb bird (*PP rr*) with a rose comb bird (*pp RR*)

Because the action of one gene at a given locus is modified by a second gene at a different locus, the two genes are said to be *complementary*.

In Sweet Peas (*Lathyrus odoratus*) colour in the flower is dependent on the presence of both dominant complementary genes *C* and *R*. If either is present in the recessive state only, then a white flower is formed. This means that there are three types of white flower *cc R—*, *C— rr* and *cc rr*. If the two white plants *cc RR* and *CC rr* are crossed, the *F1* appear coloured. Selfing the purple flowers produced gives an *F2* in the ratio of 9 coloured flower plants to 7 white (Fig. 48).

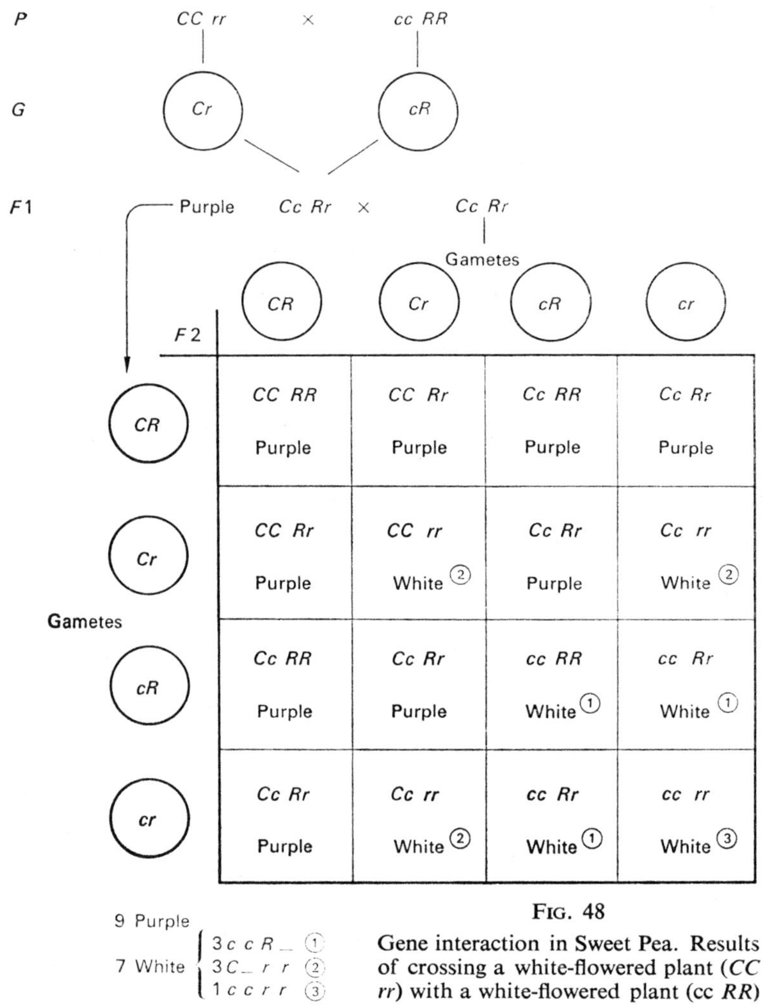

FIG. 48

Gene interaction in Sweet Pea. Results of crossing a white-flowered plant (*CC rr*) with a white-flowered plant (cc *RR*)

The 9:7 ratio is a modification of Mendel's original one: in fact the seven white plants consist of the three genotypes:

3 are *cc R*— (wl)
3 are *C*— *rr* (w2)
1 is *cc rr* (w3)

4.10 EPISTASIS

This again is a form of interaction between two non-allelic genes. However one gene, the *epistatic gene*, controls the expression of the second gene, the *hypostatic gene*. This can be illustrated by a cross in mice. The colours agouti (black hairs with yellow tips) and black are controlled by a pair of alleles *A* and *a*. Agouti is dominant. Whether or not any colour develops however depends on a second pair of alleles *C* and *c*. The presence of pigment, *C*, is dominant. The homozygous recessive *cc* produces an albino mouse. The results of a cross between a pure black mouse and an albino carrying the alleles for agouti are shown in Fig. 49.

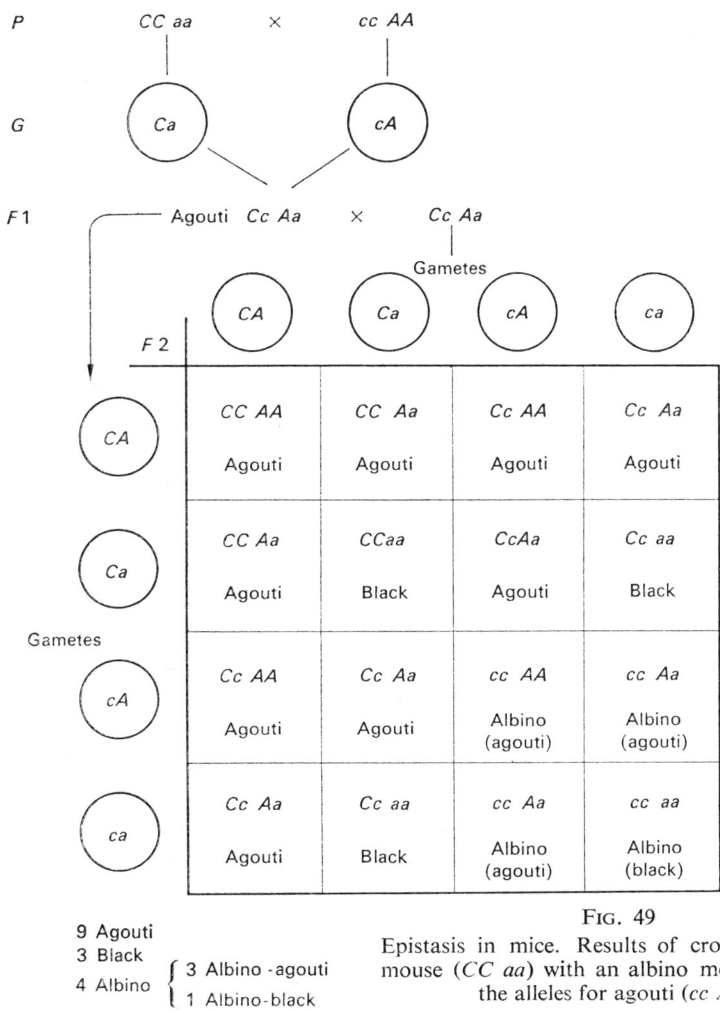

FIG. 49

Epistasis in mice. Results of crossing a black mouse (*CC aa*) with an albino mouse carrying the alleles for agouti (*cc AA*)

This again produces a modified 9:3:3:1 ratio of

9 Agouti
3 Black
4 Albino—{3—Albino agouti; 1—Albino black}

In this example, whether or not the gene *A* expresses itself depends on the state of the second gene *C*. *A* is said to be hypostatic to the epistatic gene *C*.

4.11. MULTIPLE ALLELES

The theoretical crosses that we have considered so far in this chapter have all involved a gene and its mutated allele. In a population however it is quite feasible that the original

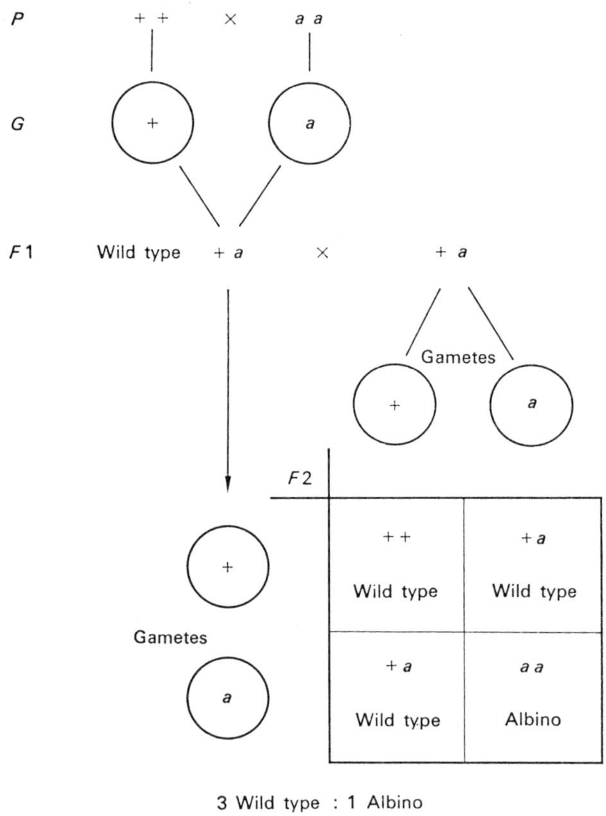

FIG. 50(a)
Multiple alleles in rabbits. Results of crossing a wild-type rabbit (++) with an albino rabbit (*aa*)

wild-type gene could have mutated more than once to give a series of possible alleles, pairs of which would be present in a given individual. An example is found in rabbits. The albino, and Himalayan which has pink eyes, a white coat and black extremities, are both mutants at the same locus of the wild-type coat colour. Wild type (+) is dominant to Himalayan (*h*) which in turn is dominant to albino (*a*). Crosses would produce the results shown in Figs. 50(a) and 50(b).

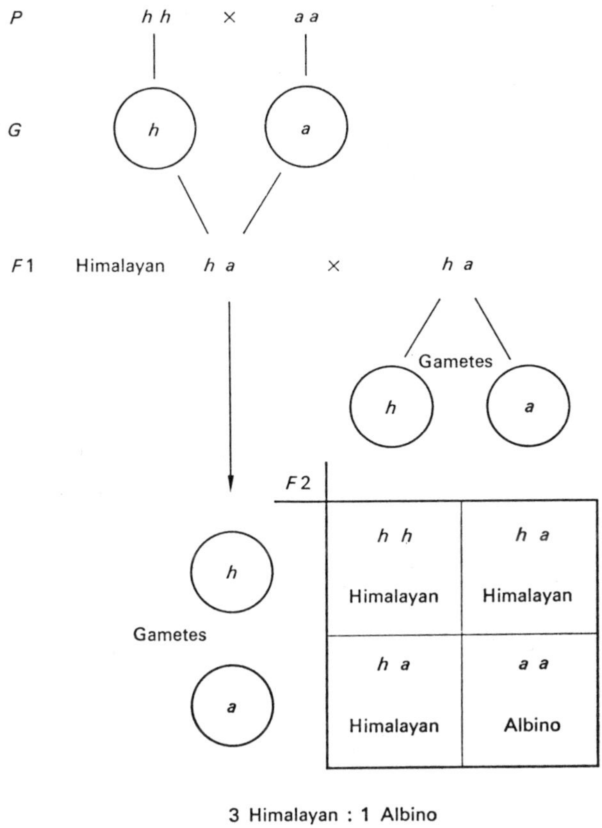

FIG. 50(b)

Multiple alleles in rabbits. Results of crossing a Himalayan rabbit (*hh*) with an albino rabbit (*aa*)

4.12. PLEIOTROPIC GENES

Up until now we have considered that a gene controls one particular character, for example the mutated gene in *Drosophila* produces vestigial wings. This 'one gene one character' notion, however, is rarely the case: instead the vast majority of genes, if not all, will affect a complex of characters. The gene for vestigial wing will also affect less noticeably the body size, legs and so on. This phenomenon is termed *pleiotropism*.

An excellent example of this multiple effect of a single gene is the mutated gene, grey lethal, found occurring in mice. This gene affects the animal's coat colour; it also upsets the development of the skeleton. In the normal development of the skeleton much remodelling takes place, with reabsorption and redeposition of bone occurring. The mutant mouse is unable to reabsorb the bone; deposition of new bone continues resulting in abnormal thickening of the bones. The marrow cavities which usually develop later by reabsorption of bone material are unable to form. The marrow is diffusely spread through the spongy bone. Because the jawbones are over thickened the teeth are unable to erupt, indeed the incisors are forced to grow backwards. The encaged bases of these teeth then compress nerves and blood vessels. The resulting neuralgia and starvation, due to their inability to feed after weaning, causes the death of the animal. Thus one gene has produced a whole 'pedigree' of effects.

4.13. CHI-SQUARED TEST

When dealing with Mendelian inheritance and with modifications of the ratios, we are concerned with theoretical ratios. In practice large numbers of offspring are often obtained and one is faced with the problem 'are the numbers obtained in practice close enough to the expected theoretical numbers to be taken as valid?' or, rephrased 'is the difference between the numbers obtained and the expected numbers a difference due to chance, or is it a real difference?'

Consider the result of a *Drosophila* cross where the expected ratio was 1:1. The number of offspring was 426 of which 228 were wild-type flies, the remaining 198 were vestigial types, instead of the expected 213:213.

To determine whether the difference is due to chance or if it is a real difference and something is wrong with the 1:1 ratio, the chi squared (X^2) test is used. For each phenotype the difference between the observed numbers and expected numbers is obtained. This difference is squared. The square of the difference is then divided by the expected number. This is repeated for each class, or phenotype, and the figures obtained are totalled. The sum is the value of X^2.

$$X^2 = \text{sum of } \left(\frac{(\text{Difference})^2}{\text{Expected number}}\right)$$

$$X^2 = \text{sum of } \left(\frac{(\text{observed numbers} - \text{expected numbers})^2}{\text{Expected numbers}}\right)$$

In the above example the differences between the observed and expected are:

		Observed	Expected	Difference
(i)	Wild type	228	213	15
(ii)	Vestigial	198	213	-15

(i) $\dfrac{(\text{Difference})^2}{\text{Expected}} = \dfrac{(15)^2}{213} = \dfrac{225}{213} = 1{\cdot}06$

(ii) $\dfrac{(\text{Difference})^2}{\text{Expected}} = \dfrac{(-15)^2}{213} = \dfrac{225}{213} = 1{\cdot}06$

$$X^2 = \text{sum of } \left(\frac{(\text{Difference})^2}{\text{Expected}}\right)$$

$$X^2 = 1{\cdot}06 + 1{\cdot}06$$

$$X^2 = 2{\cdot}12$$

This figure for chi-squared now has to be compared with a probability table prepared to show how often in 100 cases the given value (X^2) could have been produced by chance, see Table 3 below.

TABLE 3

Number of times in 100 that chance alone could have accounted for the difference	99	95	90	70	50	30	10	5	1	0·1
X^2 for two classes	0·0002	0·004	0·02	0·15	0·46	1·07	2·71	3·84	6·64	10·8
X^2 for three classes	0·02	0·103	0·21	0·71	1·39	2·41	4·61	5·99	9·21	13·8
X^2 for four classes	0·115	0·352	0·58	1·42	2·37	3·67	6·25	7·82	11·34	16·3

If we read off X^2 for crosses involving two groups of phenotype we find that 2·12 lies between 10 and 30. This means that the difference in the observed numbers could have occurred by chance alone between 10 and 30 times in every 100 experiments. When X^2 is above 5 times in 100 it is said not to be significant, that is the difference was probably produced by chance alone.

Consider a second example. In an experiment involving tomato plants the *F2* seedlings of a monohybrid cross were expected in the ratio of 3 green-leafed plants to every one yellow-leafed plant. 932 seedlings were grown of which 760 were green and 172 were yellow instead of the expected 699 green and 233 yellow.

$$X^2 = \text{sum of } \left(\frac{(\text{Difference})^2}{\text{Expected numbers}}\right)$$

$$X^2 = \left(\frac{(760-699)^2}{699} + \frac{(172-233)^2}{233}\right)$$

$$X^2 = \frac{(61)^2}{699} + \frac{(-61)^2}{233}$$

$$X^2 = \frac{3721}{699} + \frac{3721}{233}$$

$$X^2 = 5{\cdot}31 + 15{\cdot}9$$

$$X^2 = 21{\cdot}21$$

This figure for X^2 is now compared with the table for two classes of phenotype. The probability lies well below one. This means that the result could have been produced by chance alone less than once in every 100 experiments. This probability lies well below the accepted figure of 5 times in 100 and the difference is said to be significant. This means that the results do not show the 3:1 ratio expected and the cross must now be re-examined for a cause: in this case the yellow plants are not as viable as the green plants.

Crosses which produce three or four groups of offspring have to be compared with other rows in the table showing X^2 for three or four classes.

Plate V shows an *F2* maize cob from a monohybrid cross involving purple (*R*) and yellow (*r*) alleles. The *F2* cob is expected to show a 3:1 ratio of purple to yellow kernels. Score as many of the kernels as possible and, using the chi-squared test as outlined above, determine whether the expected ratio has been fulfilled.

5
Non-Mendelian inheritance

5.1. GENE LINKAGE

AFTER consideration of several dyhybrid crosses in Chapter 4, it should now be clear that the emergence of a 9:3:3:1 ratio in the *F2* generation depends upon the two types of gene assorting independently. Thus if we imagine an *F1* cross: *AaBb* × *AaBb*, the 9:3:3:1 ratio in the *F2* will only appear if each *F1* organism produces the four types of gamete: *AB*, *Ab*, *aB* and *ab*, in equal numbers. Since the publication of Mendel's work a large number of exceptions to this rule of independent assortment have been described. Like all exceptions they have thrown a good deal of light on the rule itself.

In 1906 Bateson and Punnett described such an exception in the Sweet Pea. They crossed a purple-flowered variety with long pollen grains against a red-flowered variety with short pollen grains. The *F1*, were all purple flowered with long pollen grains, thus these two alleles were dominant. On selfing the *F1* they found that the *F2* phenotypes were not in the expected ratio of 9:3:3:1. They found the *F2* contained an excess of purple/long and red/short types, that is, *parental types*. There was, of course, a corresponding deficiency of purple/short and red/long types, that is, *recombinant types*. If the two genes are considered separately, however, it was found that each segregated in the expected 3:1 ratio in the *F2*. This indicates that these two characteristics are controlled by two completely distinct genes.

T. H. Morgan made a cross between wild-type *Drosophila*, and flies with black bodies and vestigial wings. In the *F1* generation all the flies were phenotypically wild type, indicating that the alleles for black body and vestigial wing were recessive. He then testcrossed this *F1* against the double recessive flies of the *P* generation. One would expect such a testcross to give rise to the following types of fly in equal proportions:

Phenotype	*Expected ratio*	*Numbers obtained*
Wild type	1	586
Normal body, vestigial wing	1	106
Black body, normal wing	1	111
Black body, vestigial wing	1	465

Thus, as in the previous example, we have an excess of parental types, and a deficiency of recombinant types.

This tendency of parental combinations of genes to remain together, and the corresponding infrequency of new combinations, is referred to as *linkage*. In a testcross, such as the one just described, if the proportion of parental types exceeds 50%, then the two genes are said to be *linked*. The independent assortment of alleles depends upon the arbitrary separation of members of homologous chromosome pairs at meiosis. If, however, genes are situated on the same chromosome, then it is clearly impossible for alleles of those genes to assort independently, since they are physically linked together. We also know that in any organism there are many more genes than chromosomes, therefore all chromosomes must carry a large number of genes. Thus in any organism there will be groups of genes which will not assort independently, and crosses involving these genes will not give rise to classical Mendelian ratios. Although this is the derivation for the title of this chapter, it should be pointed out that study of linked genes has in no way invalidated Mendel's work, in fact, rather the contrary.

Let us now return to our example of linkage in *Drosophila*, and show the effect of a physical linkage of genes,

The *F1* generation is heterozygous for both alleles, but the wild-type alleles are linked together on the same chromosome, as are the two recessive alleles. There cannot, therefore, be independent assortment of these alleles during gamete formation, and, in theory, only two types of gamete will be produced. These gametes will give rise only to parental types when crossed with the double recessive.

5.2. CROSSING OVER

The above account of linkage would lead us to believe that, where two genes are linked together, crosses involving different alleles of such genes will only give rise to parental types. In the two examples quoted, however, we have seen that recombinant types are produced, albeit in frequencies lower than would be expected in a normal dihybrid cross. What is more the frequency of recombinants produced in any particular cross involving linked genes remains fairly constant. We must therefore look for some mechanism whereby linked alleles are 'recombined' in a relatively organised fashion.

Towards the end of prophase of the meiotic division by which the gametes are formed, the homologous chromosomes appear to be joined together at one or more points termed chiasmata (see Plate IIId). At this stage of the division it is important to remember that each chromosome consists of two identical sister chromatids, themselves held together at the centromere. We must also bear in mind that alleles of a gene occupy corresponding positions along the length of members of a pair of homologous chromosomes. There is now substantial evidence to suggest that at the chiasmata breakages occur at corresponding points along *non-sister* chromatids, Fig. 52(b).

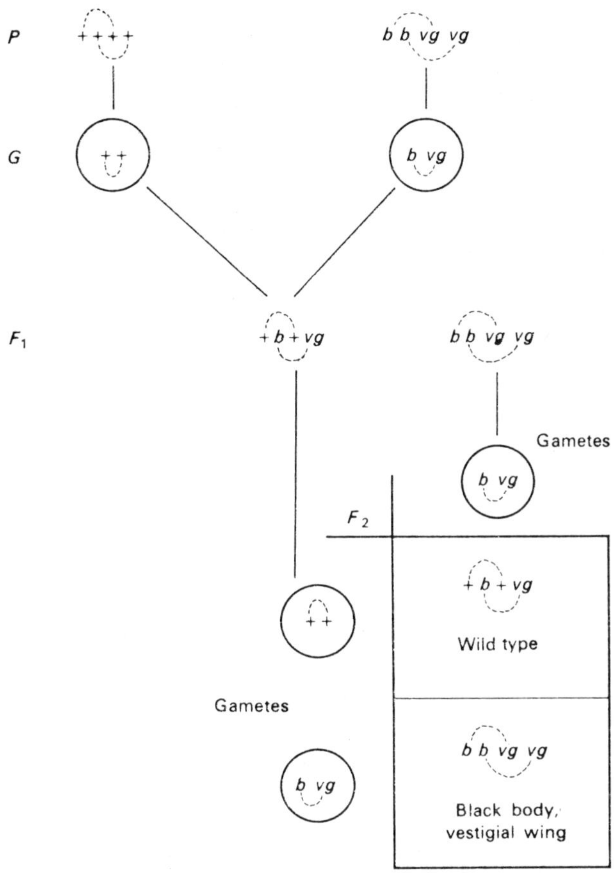

FIG. 51

Drosophila. A cross between wild-type flies and flies homozygous for black body (*b*) and vestigial wing (*vg*). Linkage of alleles is indicated by a dotted line. The *F1* is testcrossed to the homozygous recessive.

Rejoining of the chromatids then takes place, but this may be between non-sister chromatids, Fig. 52(c). During anaphase I of meiosis the homologous chromosomes separate to opposite poles of the dividing cell, Fig. 52(d). This is rapidly followed by the second division of meiosis in which the centromere in each chromosome divides and the sister chromatids move to opposite poles in each case. As Fig. 52(e) shows two of the gametes produced will contain the parental type of chromosome, but two will contain recombinant chromosomes made up of part of each parental type. This process is termed *crossing over*.

Cytological evidence for crossing over was produced in 1931 by Stern. He found *Drosophila* which contained irregular *X* (sex) chromosomes. One of these was an *X*

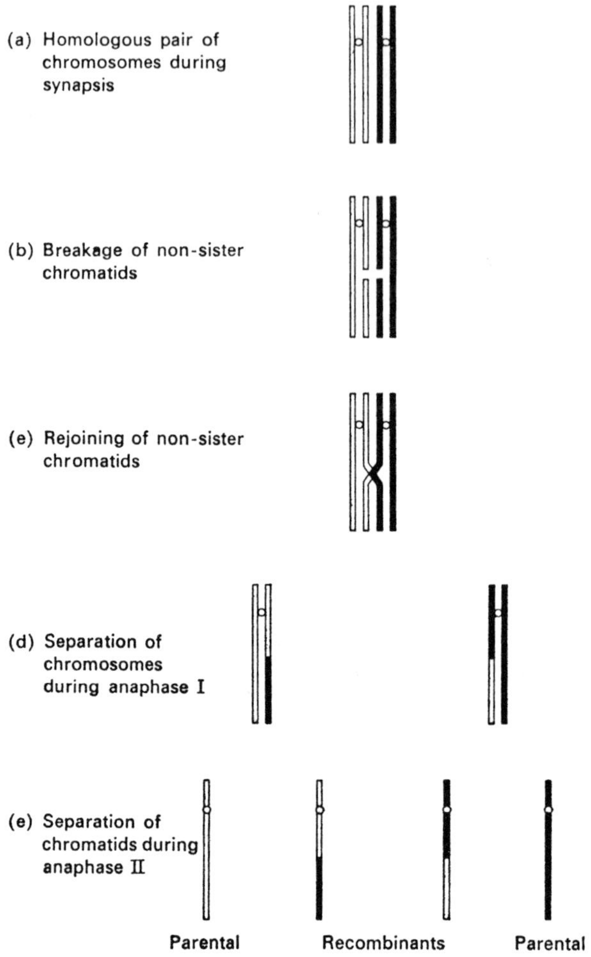

FIG. 52
Crossing over during meiosis

chromosome to which a portion of a *Y* chromosome was attached. In the other the *X* chromosome was broken and part of it had become attached to one of the autosomes. By crossing these two strains he obtained females in which both *X* chromosomes were abnormal, one of each type. Figure 53 shows the results of a cross between a female of this type which was heterozygous for carnation eye (recessive) and bar eye (dominant), and a carnation male. Both of these genes are carried on the *X* chromosome. Microscopic examination of the chromosomes of the progeny showed that flies with recombinant phenotypes also showed changes in the appearance of the chromosomes. Such changes could only have been brought about by crossing over taking place during egg formation in the female. The crossing over must have occurred between the carnation and bar genes. In this case the small portion of the *Y* chromosome attached to the *X* chromosome acted as a visible indication of crossing over. Normally we depend upon the appearance of recombinants in the progeny to indicate crossing over.

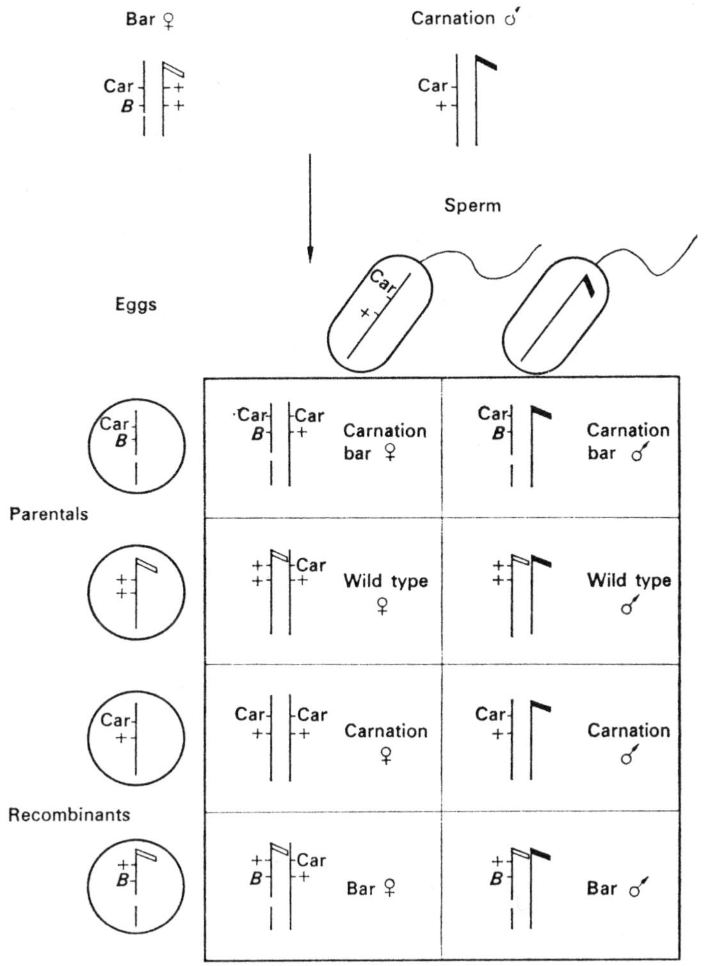

FIG. 53
Stern's experiment giving cytological proof of crossing over in *Drosophila*

In the examples quoted so far the situation is somewhat complicated by the dominance of one allele. The effects of crossing over can be seen far more rapidly by observing the life cycle of an organism which shows a predominant haploid phase where the presence of only one allele means that dominance is of no importance.

5.3. CROSSING OVER IN SORDARIA FIMICOLA

Sordaria fimicola is an Ascomycete fungus in which the life cycle is entirely haploid, with the exception of a diploid zygote stage in sexual reproduction. It is homothallic and will undergo sexual reproduction without the presence of different mating strains. Although the exact details of sexual reproduction are not yet known, it seems likely that there is fusion of hyphae, and the passage of nuclei from one hypha to the other. Zygotic nuclei are formed by nuclear fusion, each being contained within a sac-like structure, the ascus. Quite a large number of these asci develop within a pigmented, flask-shaped structure, the perithecium. The zygotic nuclei undergo meiosis, each of the four cells produced then divides mitotically to give a total of eight cells within each ascus. These eight haploid cells develop a resistant wall and form ascospores.

The genetic importance of this process lies in the fact that because of the long, thin shape of the ascus, the spindles formed during the meiotic and mitotic divisions will all lie along the length of the ascus. Thus the final position of the ascospores in the ascus can be related to the position of the chromosomes during the meiotic division of the zygote. To illustrate the importance of this ordered arrangement of ascospores

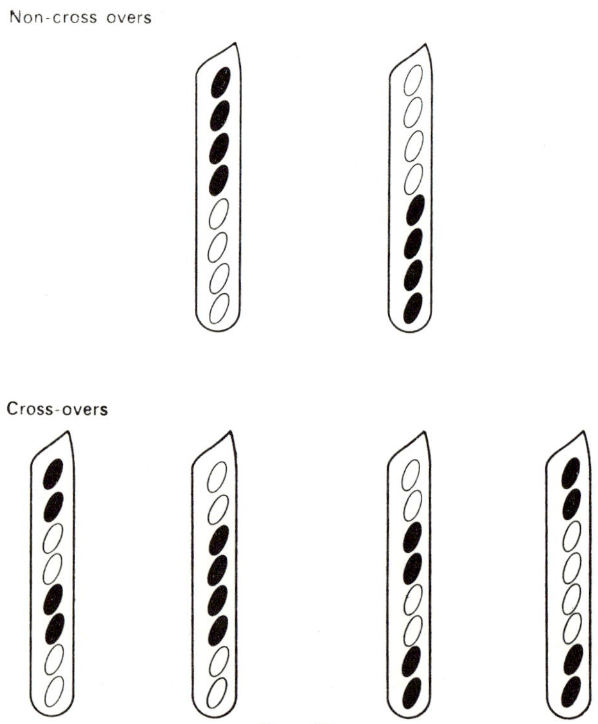

FIG. 54

Sordaria. The arrangements of ascopores produced when white and black spored strains are crossed we will take as our example a gene which controls the colour of the ascospore. Normally the spores are black, but there is an allele, giving rise to white spores, which occurs at the same locus, that is the position on the chromosome. If black and white spored strains are inoculated into the same agar plate and incubated (see Chapter 10), a large number of perithecia are formed. If the asci within a perithecium have been formed as a result of a cross between the two strains, then each ascus will contain four black and four white ascospores. However, the arrangement of the ascospores will vary, the six possible arrangements being shown in Fig. 54.

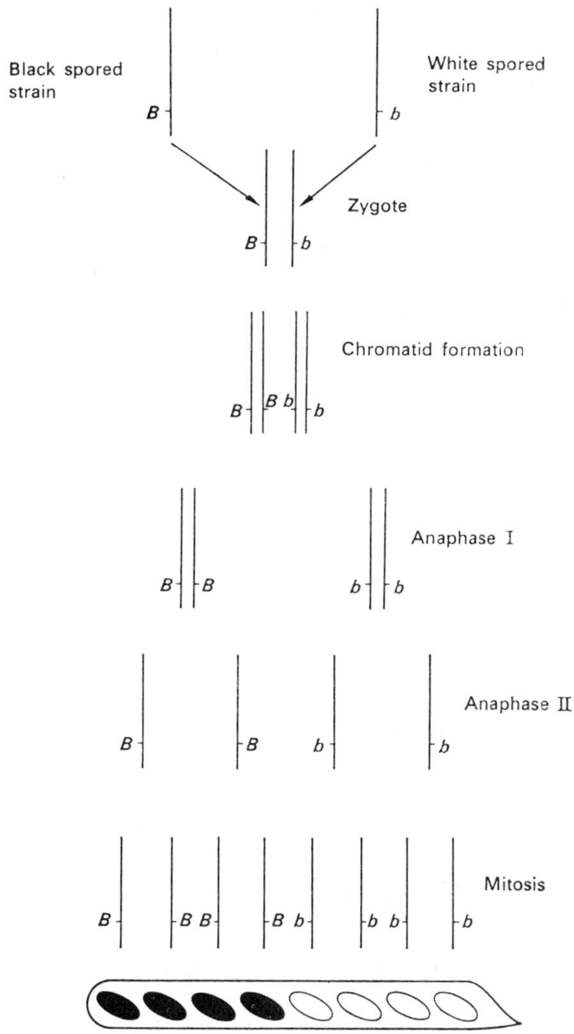

FIG. 55

Segregation of the black and white alleles during the first meiotic division

The first two possibilities have arisen by a straightforward segregation of the two alleles during the first division of meiosis. The alleles, carried on separate members of a homologous pair of chromosomes, move to opposite poles during anaphase I. The subsequent separation of chromatids during anaphase II, followed by mitotic division, gives a group of four ascospores carrying the allele for black colour, and a group of four carrying the allele for white colour (Fig. 55). The other four combinations are explained on the basis of crossing over taking place between the centromere and the locus of the gene in question. An example is given in Fig. 56, but this over-

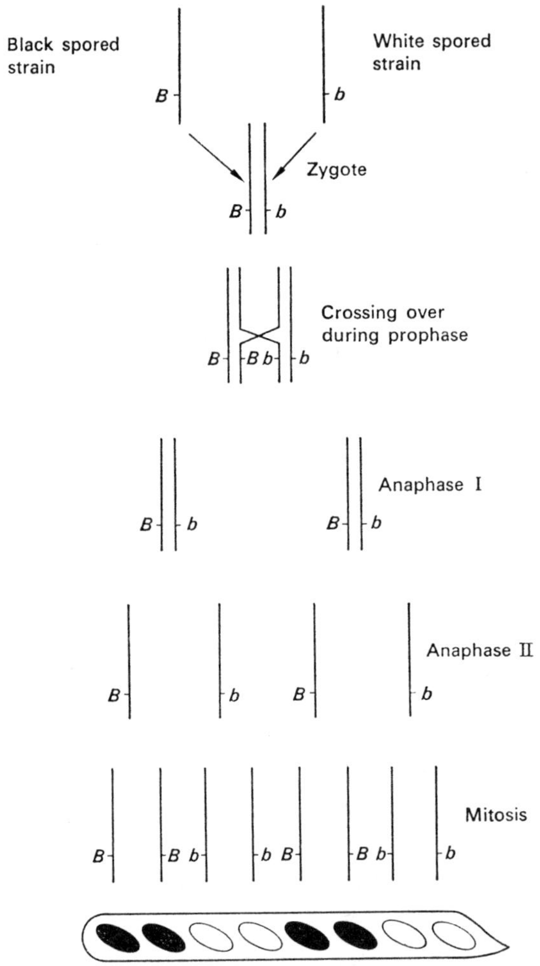

FIG. 56
Sordaria. Segregation of the black and white alleles during the second meiotic division due to crossing over

simplification is somewhat confusing in that crossing over begins as the chromosomes are coiled around each other, and there are no definite 'inner' and 'outer' chromatids as shown. The basic point is that after anaphase I each chromosome will consist of two chromatids carrying different alleles. The exact orientation of the alleles will depend upon which chromatids were involved in the crossing over. Segregation of the alleles occurs in anaphase II of meiosis and subsequent mitosis will give one of the second group of four possible arrangements shown in Fig. 54.

5.4. MAKING A CHROMOSOME MAP

In organisms where there has been an intensive study of linked genes it has been shown that the proportion of recombinants produced in any particular cross is fairly constant. This proportion is, however, different from the proportions formed in crosses involving other pairs of linked genes. We can explain this on the basis that firstly each gene has its own fixed locus on the chromosome, and secondly that crossing over is more likely to occur between more widely spaced genes. Thus if a cross involving two linked genes gives rise to a small proportion of recombinants we can assume that the genes in question are separated by only a short distance along the chromosomes, and *vice versa.* Sturtevant suggested that the percentage of recombinants could be used as a measure of the distance between genes, where 1% is equal to one map unit.

For example: crosses involving the linked genes *X* and *Y* give 5% recombinants. Therefore their positions can be represented:

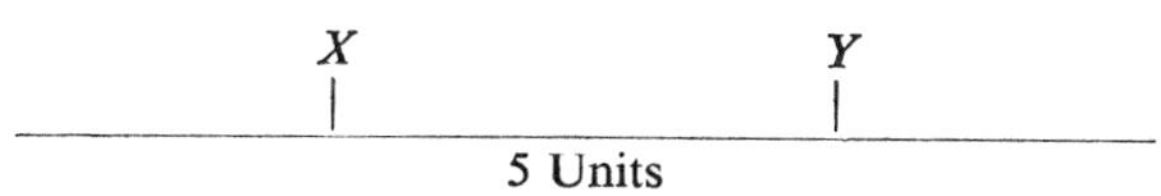

A third linked gene, *Z*, gives rise to 10% recombinants when crossed with *X*. This gives two possible positions for *Z*:

(a)

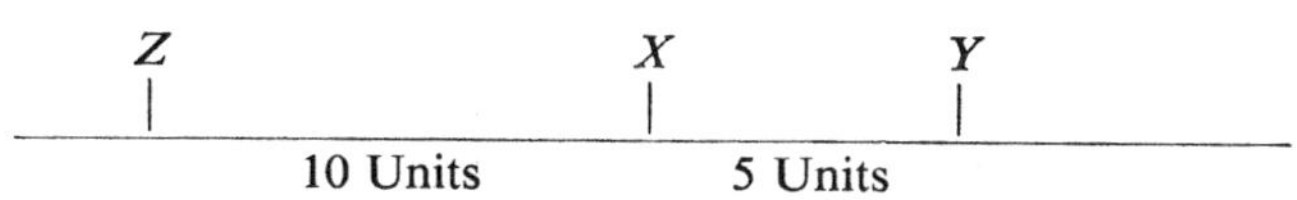

or (b)

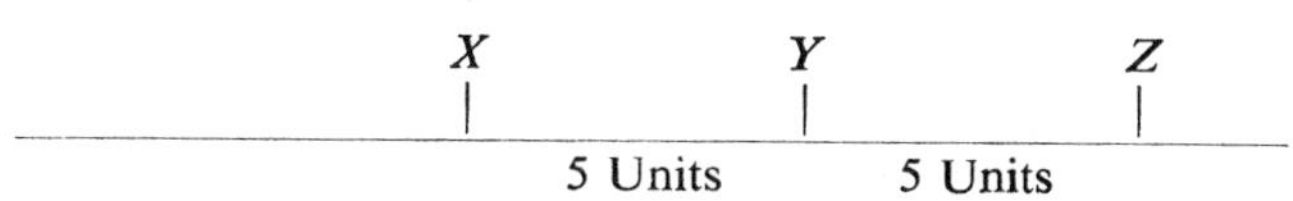

The precise position of *Z* can be determined by carrying out a cross involving *Y* and *Z*. If the gene *Z* is positioned as shown in (a) we would expect 15% recombinants, if as in (b) 5% recombinants.

By using this basic technique the positions of large numbers of genes in a variety or organisms have been fixed. As an example we shall take a cross in maize, where several pairs of alleles control the nature of the endosperm.

C—coloured *c*—colourless
Wx—non-waxy *wx*—waxy

P *CC Wx Wx* × *cc wx wx*

↓

F1 *Cc Wx wx* × *cc wx wx*
Coloured, non-waxy

F2		Number of offspring
Coloured, non-waxy	*Cc Wx wx*	2542
Coloured, waxy	*Cc wx wx*	717 }
Colourless, non-waxy	*cc Wx wx*	739 }
Colourless, waxy	*cc wx wx*	2710

The high proportion of parental types indicates that these two genes are linked. The percentage of recombinant types, that is those bracketed, is 21·7. We can convert this percentage into map units, giving the distance between the two genes.

The two genes are found on chromosomes number 9, a more detailed map of which is shown below:

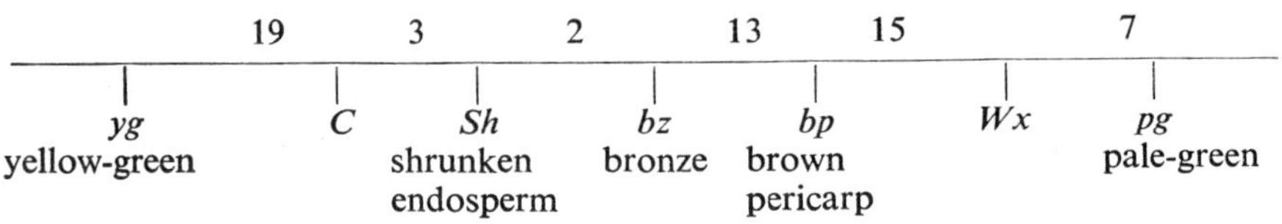

It can be seen from the map that the distance between *C* and *Wx* is in fact 33 map units. The discrepancy between this figure and the one calculated from the number of recombinants is almost certainly due to double crossing over, as shown in Fig. 57. All the chromatids resulting from this process are of the parental type and therefore the observed frequencies of recombinant types can be an underestimate of the amount of crossing over. This inaccuracy can be largely overcome by choosing genes close together on the chromosome thus greatly reducing the likelihood of a double cross over occurring between them.

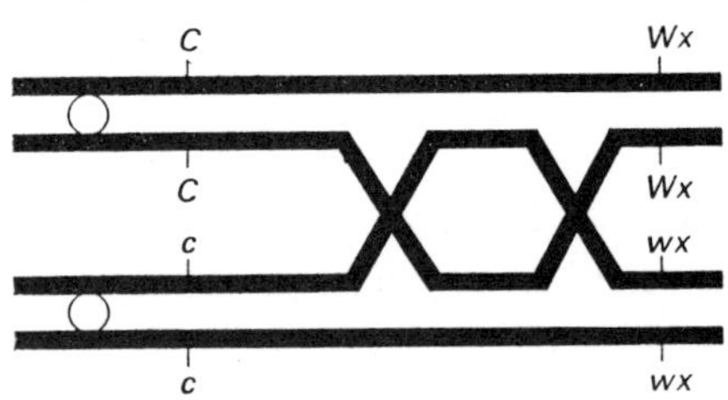

FIG. 57

How double cross overs can lead to a higher proportion of parentals than expected from map distance

5.5. SEX LINKAGE

Increasing study of linked genes has made it clear that in the inheritance of some genes there appears to be a correlation between the genes transmission and the sex of the parents and offspring. A classic example in human genetics is red-green colour-blindness, in which these two colours are, to varying degrees, indistinguishable. Study of families with a history of this anomaly demonstrates three points:

(a) Colour-blind men are far more frequent than women.
(b) Colour-blind women all have colour-blind fathers.
(c) A woman with normal vision, but who is a carrier of the recessive colour-blind gene will have sons half of whom are affected whatever the constitution of the father.

In *Drosophila* there is a recessive gene which gives rise to white, unpigmented eyes. This is termed white eye. If white-eyed females are crossed with normal males, one would expect all the *F1* to be normal. In fact all the females of the *F1* are normal, but all the males are white-eyed. In the *F2* generation the following ratios are formed:

White-eyed females	1
Normal females	1
White-eyed males	1
Normal males	1

If the reciprocal cross is made, i.e. between a normal female and a white-eyed male all the *F1*, both males and females are normal and in the *F2* the following are obtained:

Normal females	2
Normal males	1
White-eyed males	1

For an explanation of these, and other similar crosses, it is necessary to look briefly at the genetic mechanism by which the sex of an organism is determined.

5.5.1. *Sex determination*

In a large number of sexually reproducing organisms it is possible to distinguish morphological differences between the chromosomes of males and females. In *Drosophila*, for instance, there are four pairs of chromosomes, three of which are identical in males and females. In females the fourth pair are identical, and termed *X* chromosomes. In the male, however, there is one *X* chromosome only, paired with a quite distinct chromosome, the *Y* chromosome (Fig. 29). The segregation of the sex chromosomes and the possibilities of subsequent rejoining are shown in Fig. 58. This diagram also explains how the mechanism maintains a 1:1 ratio between the sexes.

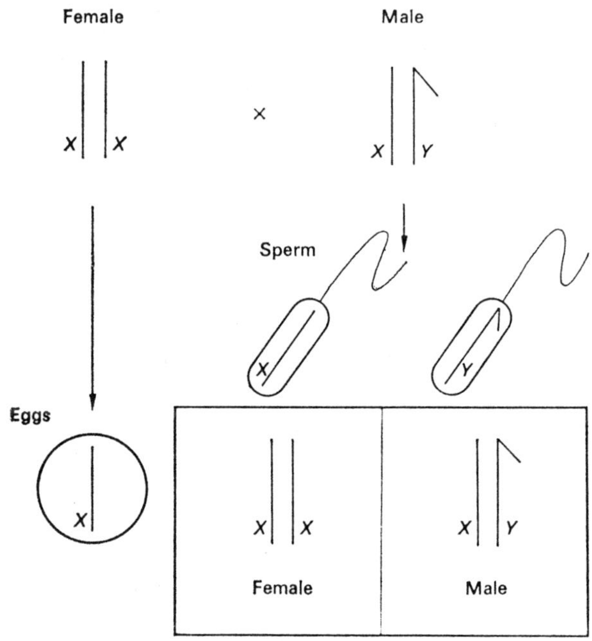

FIG. 58
The *X*—*Y* method of sex-determination

Although sex determination in man follows a superficially similar pattern in that there are recognisable *X* and *Y* chromosomes and that the female is *XX* and the male *XY*, there are basic differences. In *Drosophila* it is the proportion of *X* chromosomes to autosomes (non-sex chromosomes) which determines sex. Thus a proportion of two *X* chromosomes to three pairs of autosomes gives a female while one *X* chromosome to three pairs of autosomes gives a male. Evidence for this is found in flies which are *XO*, that is where there is only one *X* chromosome but no *Y* chromosome. Such flies are male. In man, however, it is the presence of the *Y* chromosome which determines maleness, and therefore individuals which are *XO* are female, and those which are *XXY* are males.

This type of sex determination is not universal. Some animals have no *Y* chromosome at all, and have what is described as '*XO*' sex determination, for example, *Chorthippus parallelus* (Plates IIIf, g and i). In birds the situation found in *Drosophila* is reversed in that the male has *XX* and the female *XY*. The majority of reptiles as well as some fish and amphibia seem to have the *XX*: male and *XY* or *XO*: female type of determination.

If we return now to our examples of sex-linked inheritance we can make use of our knowledge of sex determination in offering explanations.

As well as carrying genes controlling the sexual development of the individual the sex chromosomes, particularly the *X* chromosome, carry genes which affect normal body development. Such is the case in *Drosophila* where the gene controlling eye pigmentation, and its allele giving white eye are carried on the *X* chromosome. In the first cross described the female will contain two *X* chromosomes, each carrying the recessive white allele. The male will have only one *X* chromosome carrying the normal allele. The *Y* chromosome carries no genes affecting eye colour. The cross, diagrammed

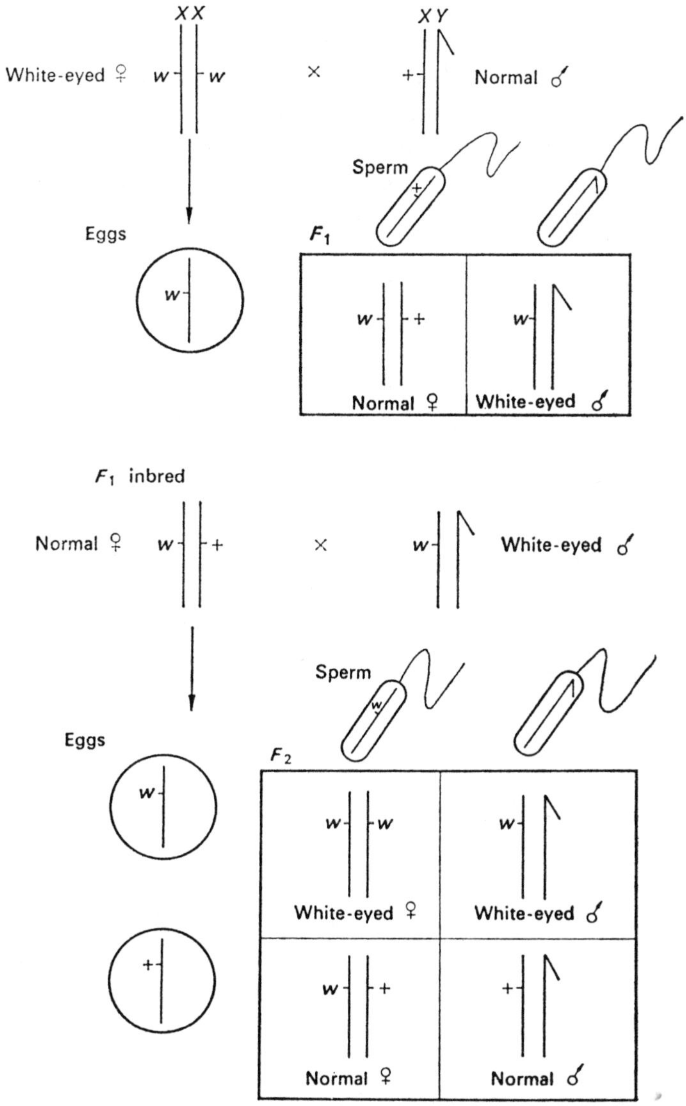

FIG. 59
The sex-linked inheritance of white eye in *Drosophila*

in Fig. 59, produces an *F1* of normal females and white-eyed males. The diagram also explains the formation of the given ratio in the *F2*. Using the same assumption, the ratios obtained in the reciprocal cross are explained; as an exercise try and diagram this cross. Note that the male's *X* chromosome must always be carried into female offspring and that because the *Y* chromosome does not carry any relevant genes the males need only one recessive allele, on the *X* chromosome, for the trait to be expressed in the phenotype.

The same argument is used in explanation of the inheritance of red-green colour-blindness in man. The condition is caused by a recessive allele carried on the *X* chromosome. It is far more common in males because only one recessive allele is necessary for it to be expressed in the phenotype, the *Y* chromosome carries no corresponding allele. For women to be homozygous for the recessive allele they must inherit it on the *X* chromosomes from both parents, and therefore their fathers must be colour-blind. A woman heterozygous for the recessive allele will produce gametes containing the normal and colour-blind alleles in equal numbers, and therefore, whatever the genotype of the father, has a 50:50 chance of producing colour-blind sons.

Before leaving sex linkage mention must be made of the disease haemophilia. In this disease the blood fails to clot normally and any small cut or bruise can lead to very extensive bleeding. The disease is caused by the presence of a recessive gene carried on the *X* chromosome, which therefore behaves in much the same way as red-green colour-blindness. Since males have only one *X* chromosome they are far more likely to exhibit haemophilia than women. The disease was introduced into several of the European royal houses by offspring of Queen Victoria. A family pedigree showing some of the lines of inheritance from Queen Victoria is shown in Fig. 60. There is no evidence of the disease in ancestors of Queen Victoria and it has been suggested that a spontaneous mutation took place, changing one of the normal alleles on the *X* chromosomes in Queen Victoria to a recessive allele causing haemophilia.

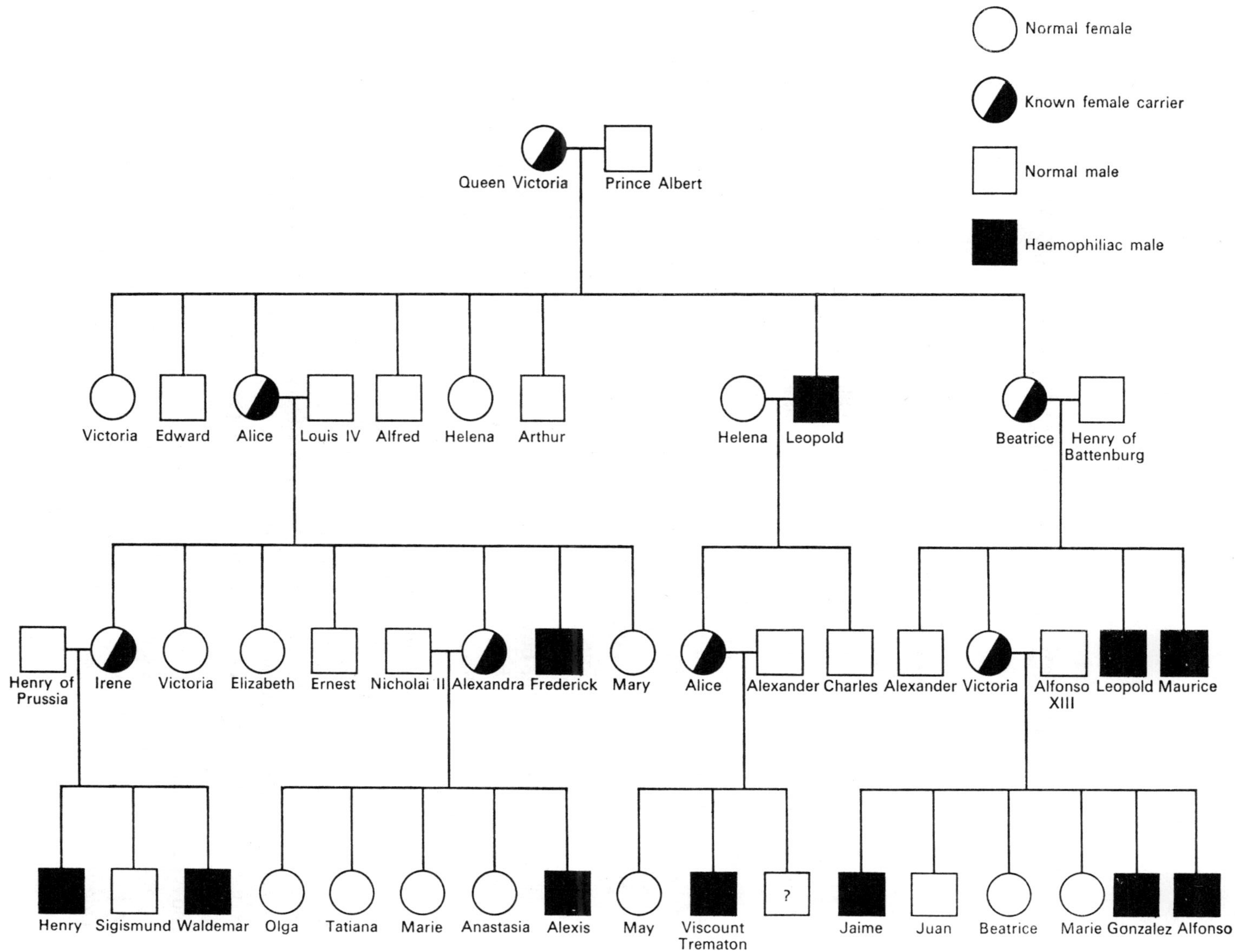

FIG. 60

A pedigree to show the incidence of haemophilia among descendants of Queen Victoria

6

The sources of phenotypic variation

6.1. CONTINUOUS AND DISCONTINUOUS VARIATION

THE variability that we have considered so far has been characterised by the fact that it gives rise to groups of organisms quite distinct from one another with respect to certain features. The normal wings in *Drosophila* are quite different from the vestigial form. The fruit of the tomato were either round or elongate. If a collection of people were asked to taste phenyl-thio-carbamide (P.T.C.) they would form two distinct groups, those who were able to taste the chemical and those who were not (Chapters 8 and 10). When Mendel examined his pea plants he found they were either tall or short. This type of variation, with no overlap between types, is called *discontinuous variation.*

We find a different situation, however, if we consider, for example, height in man. Figure 61 shows the distribution of height in 11–12-year-old Salford boys. Here instead of distinct groups as in the above examples there is an overlapping variation in their height. The majority of the boys are of 'average' height, some are a little taller or shorter, a few are extreme. This type of variation is called *continuous variation* and probably the bulk of variations between individuals of the same species are of this type. Some of the variations will be due to differences in the environment but the underlying pattern is genetically determined.

At first sight this type of variation does not appear in any way akin to the Mendelian inheritance we have previously considered. The explanation lies in the fact that characters showing this type of inheritance are controlled by a great many genes at different loci. Ignoring for the time being the effects of the environment, the overall phenotype is derived from the cumulative effects of all of the contributing genes. The example quoted above, that of height in man, is governed by an unknown number of genes—but for simplicity regard two genes a and b as being responsible. The genes a and b have alleles a^t and b^t, each of which if present increases the height of the individual by one factor. The alleles a and b will be genes for 'shortness' and a^t and b^t genes for 'tallness'. Consider then a cross between the two extremes, very short $aabb$ and very tall $a^ta^tb^tb^t$. The *F1* generation a^tab^tb will be of medium height. The results of crossing

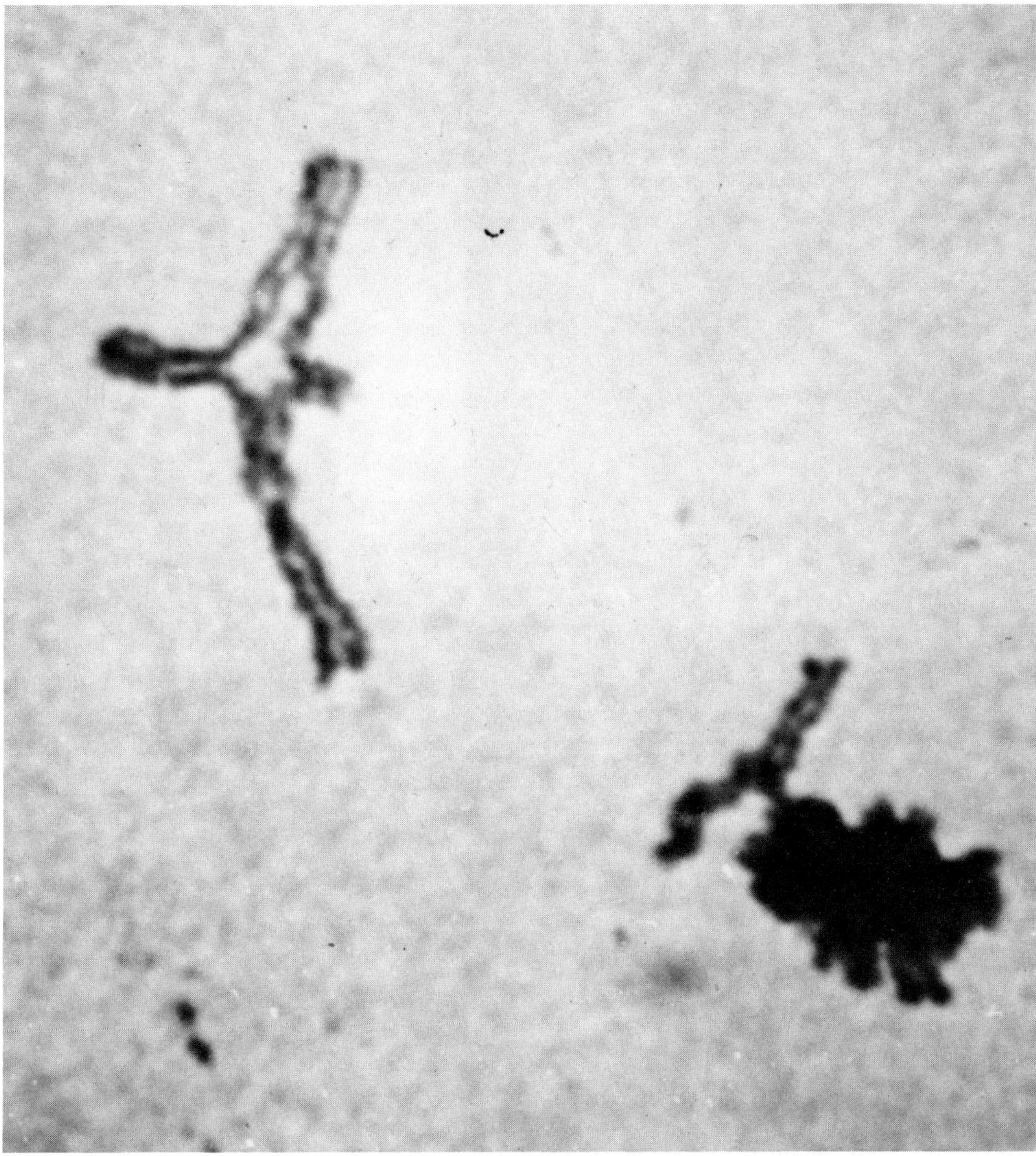

PLATE VI
The Pachytene Cross (see Fig. 67(ii))

a

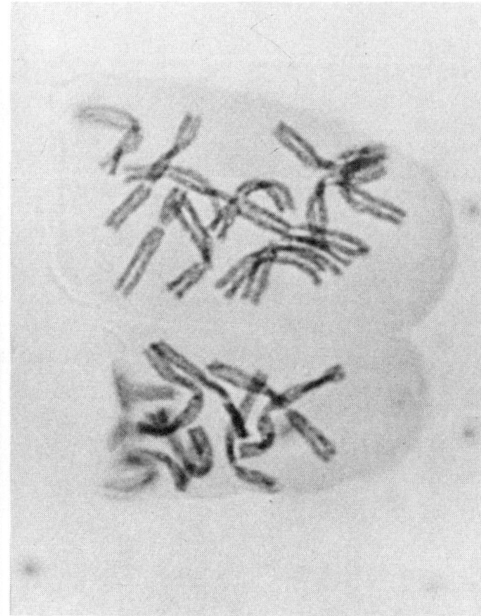

b

c

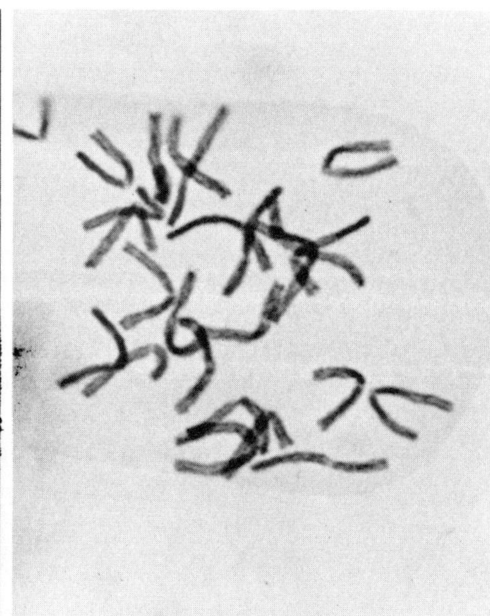

d

PLATE VII

a *T. brevicaulis*, vegetative structure
b *T. brevicaulis*, two cells in pro-metaphase, $2n = 12$
c *T. virginiana*, vegetative structure
d *T. virginiana*, single cell in pro-metaphase, $4n = 24$

Note the chromosomes are the same size in the two species but the tetraploid *T. virginiana* shows an increase in cell size.

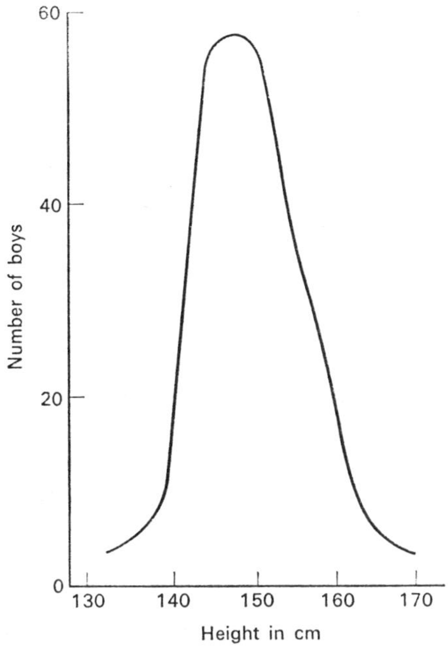

FIG. 61
The distribution of height in 11–12-year-old boys

heterozygotes are shown in Fig. 62, and if the number of 'factors' present in each genotype are totalled and then the distribution of height in the *F2* population plotted graphically the result shown in Fig. 63 compares favourably with the actual distribution shown in Fig. 61. With an increase in the number of contributing genes the distinction between the groups becomes less marked, and finally the environmental factors give a 'smooth' curve.

6.2. THE EFFECT OF THE ENVIRONMENT ON THE PHENOTYPE

The phenotype of an organism will depend not only on the basic plan set out by the genotype but also on the modifying effects of the environment. A mutant form of tobacco produces seedlings which are devoid of chlorophyll. A gene involved in the pathway which synthesises the chlorophyll has mutated thus blocking the chain of events, and resulting in a chlorotic plant. A similar result, however, can be obtained by altering the environment. Normal plants grown in the absence of light also fail to form chlorophyll. Plants grown in a medium deficient in iron or magnesium once again are unable to form the pigment. The same modification to the phenotype has been produced by a number of environmental factors.

Mendel's tall and dwarf peas differed by one mutated gene. If the dwarf peas are

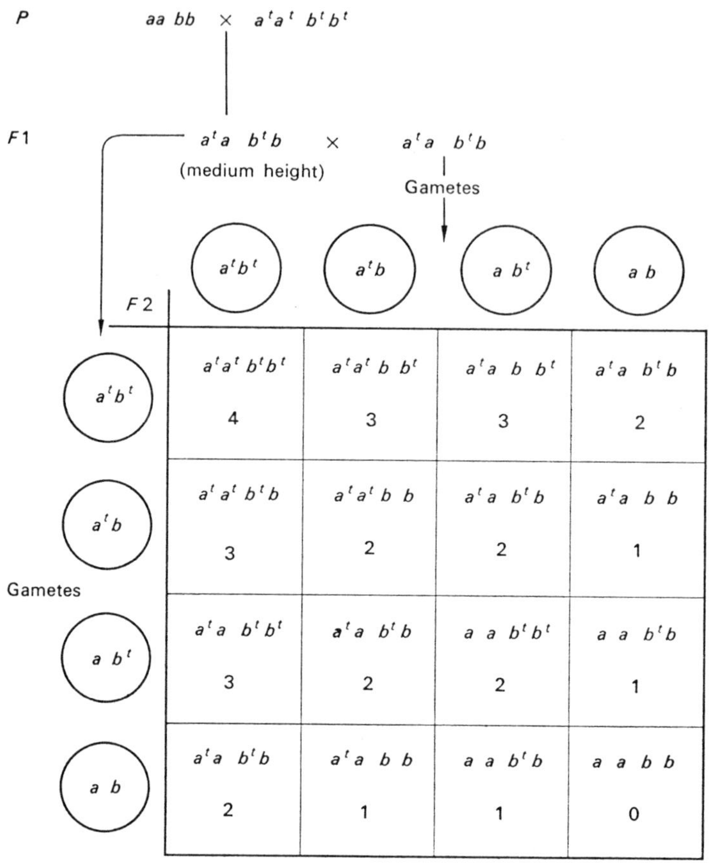

FIG. 62

Theoretical cross between a short individual (*aa bb*) and a tall individual ($a^ta^t\ b^tb^t$)

treated with gibberellic acid, a plant hormone, growth is resumed and plants indistinguishable from the tall strain are formed. Addition of an extra factor into the environment has completely altered the phenotype. Presumably the original dwarfness of the pea is due to a mutation in the combination of genes concerned with the synthesis of gibberellic acid.

A similar phenomenon occurs naturally in insects. A bee colony consists of three phenotypes, the male drones, the worker bees and the queens. The drones develop from unfertilised eggs laid by the queen bee, but both the queens and worker bees develop from fertilised eggs, in other words they have a similar amount of genetic material. Phenotypically, however, they are quite different, the workers are smaller than the queens, their mouthparts are larger, and they are sterile. The differences are due to the food

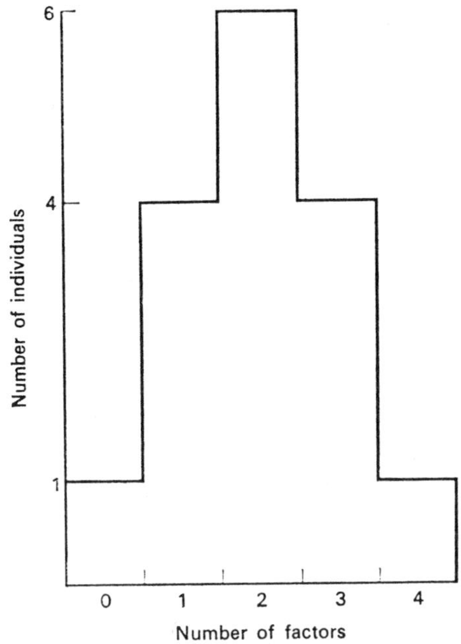

FIG. 63

The distribution of factors (t) for increased height from the $F2$ of a theoretical cross

administered to the insects during their larval period. Up until three days after emerging from the eggs all of the larvae are fed on a secretion, formed by glands on the heads of the worker bees, called 'royal jelly'. At about three days the prospective workers are switched to a diet of honey and pollen. The prospective queen larvae in the meantime continue with the royal jelly. This secretion contains a high concentration of protein and its effect is to stimulate the formation and maturation of the female reproductive system. The end result is a fertile queen. Should the queen die, and not have a replacement, the workers will immediately feed excessive amounts of royal jelly to the youngest available 'worker' larva. This larva now develops into a mature queen, instead of the sterile worker it was previously destined to become.

An example already discussed (Chapter 2) is the action of lactose on *E. coli.* This carbohydrate diffuses into the bacterial cell stimulating the DNA/RNA system to produce β galactosidase and permease, enzymes which are not normally present in such large quantities if the lactose is missing from the environment.

Most of the cases quoted above are very obvious effects of the environment on the phenotype, but many less noticeable effects are taken for granted. The rich diet producing the heavier man, a deficiency of iron resulting in anaemia, the heavy crop of apples produced one year by favourable spring conditions: all are modifications brought about by the environment.

Variations in the phenotypes are therefore the product of an interaction between the genotype and the environment. The genotype states the basic plan, the environment adds the finishing touches. It is only the genotype, however, that is inherited and we must now turn our attention to the source of genetic variations.

6.3. GENE MUTATION

The gene is the basic unit of inheritance (see Chapter 2). This complex sequence of bases on the sugar-phosphate backbone governs the formation of proteins within the cell. Many of these proteins will be enzymes which in turn control the cell's metabolism and thus its phenotype. In the gene as with any other complex piece of machinery its efficiency depends on each part of its structure. Any change, either addition or deletion to the base sequence means that the information passed to the ribosomes will be incorrect and a protein with an abnormal primary structure would be formed. The activity of the enzyme is dependent on the structure of the active centre. An incorrect primary structure might interfere with the final twisting into the tertiary form and thus produce a distorted active centre. Any change in the molecular structure of the gene would result in the loss of a particular enzyme and the subsequent blocking of a chain of reactions. An example is that of albinism, where mutation of the gene which dictates the formation of tyrosinase, results in a blocking of the reaction which forms the pigment melanin.

Although the genetic structure is remarkably stable and resistant to accidental alteration, changes do occur. These naturally occurring changes in the molecular structure of the gene are called *spontaneous gene mutations.*

When an allele mutates it produces a heterozygous condition. Usually the mutant allele is recessive and its effects are masked by the partner allele. Thus in a population there will be a reservoir of recessive mutated genes, all potential contributors to variation. Many of the mutations are harmful to the organism but their effects are masked by heterozygosis. Inbreeding will encourage the formation of homozygotes and increase the probability of such recessive mutations appearing (see Chapter 7).

The actual structure of the gene is so complex that mutations may occur at a great many points within the gene; thus providing many variations on the same theme and resulting in a multiple allelic condition. An example showing this mutation at several points within the gene is found in *Drosophila*. The *X* chromosome carries a gene which determines eye colour. Mutation of this gene has produced the white-eye condition (Chapter 5). However, further mutations at the same locus have occurred to give a range of eye colours from 'blood' and 'coral' (the reds), through pink, 'eosin' and 'cherry' to cream, 'ivory', as well as white. Each is thought to have arisen by an alteration of the molecular structure at different points within the same gene.

Gene mutations which are inherited must have originally occurred either during gametogenesis or in the cells which will give rise to the germ tissue. In this way they will be passed on to future generations. Mutations may also take place in the genetic material of the body cells. This *somatic mutation* will result in a mosaic condition where the mutated gene is present in some but not all of the body cells. Such cases are relatively common in plants, one well-known example in *Pelargonium* is the sport, or chimaera, having the variegated leaf. The commonest type has a white margin and green centre; a less common form has a green margin and white centre.

6.4. CHROMOSOME MUTATIONS

Besides mutations at the molecular level within the gene, variation can be produced by an alteration involving the chromosomes, either in part, as a whole or in complete sets.

6.4.1. *Changes involving parts of the chromosomes*

(a) *Inversions*. This type of mutation involves a section of a chromosome becoming detached and replaced in an inverted position. At first this sounds hardly a credible situation but it becomes more acceptable if the chromosome was looped along its length. An inversion would occur if the chromosome became segmented at the points of intersection, and the detached ends rejoined with the incorrect parts of the segment (see Fig. 64). Such inversions do not remove any genetic information from the cell and

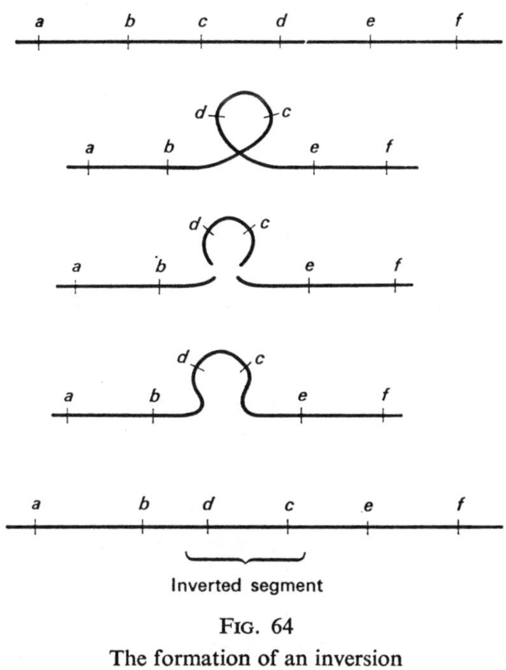

FIG. 64
The formation of an inversion

some inversions have little effect on the organism. However complications will arise during the 'gene opposite gene' pairing of synapsis. For this arrangement to take place the chromosome carrying the inverted segment will have to form a loop, with the homologue lined up along the outside (Fig. 65). If chiasma formed between chroma-

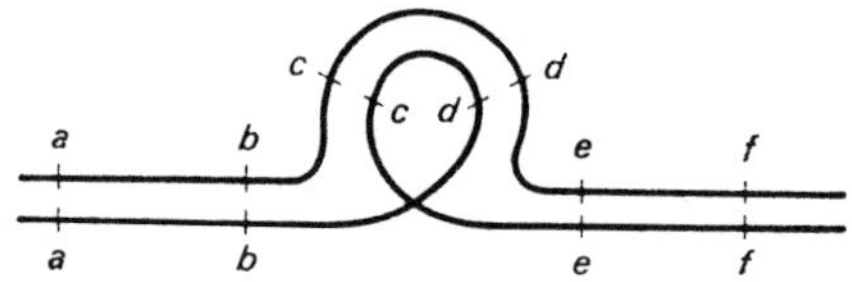

FIG. 65
Gene by gene pairing by the formation of an inversion loop

tids inside the loop, both of the products would contain duplicated genes, and both would carry deficiencies (Fig. 66). Such products as these are likely to be inviable. In other words the only products to survive would be those which did not attempt chiasma formation, this in turn means that inversions tend to suppress chiasma formation, and

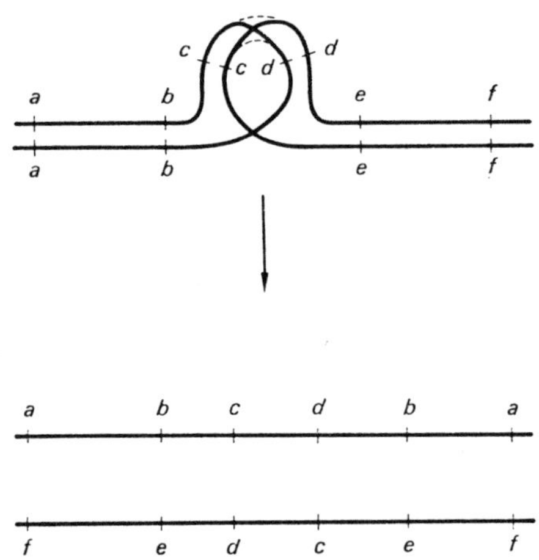

Both products have duplications and deletions

FIG. 66
Chiasma formation inside an inversion loop

this phenomenon has been used to prove the existence of inversions. Direct evidence of inversions can be obtained cytologically by examining the giant chromosomes in the salivary glands of *Drosophila* (see Chapter 10 for the method) where inversions, if present, can be seen as an inversion loop.

(b) *Translocations.* Translocations involve the exchange of segments between non-homologous chromosomes. This means that new sets of genes will behave as though they are linked together (Chapter 5), and because no genetic material is lost the products may be viable. As with inversions complications arise during meiosis; at pachytene the homologues pair off, but with a chromosome carrying a translocated segment it has to synapse with two chromosomes. The bulk of it will lie alongside its original homologue; the translocated section will synapse with the identical portion of its previous homologue. This type of pairing will also occur with the second chromosome involved and a characteristic Pachytene Cross is formed. As meiosis passes from pachytene through diplotene to diakinesis the homologues repel each other, and the cross-configuration changes to form a translocation complex in the shape of a ring. (See Fig. 67 and Plate VI.)

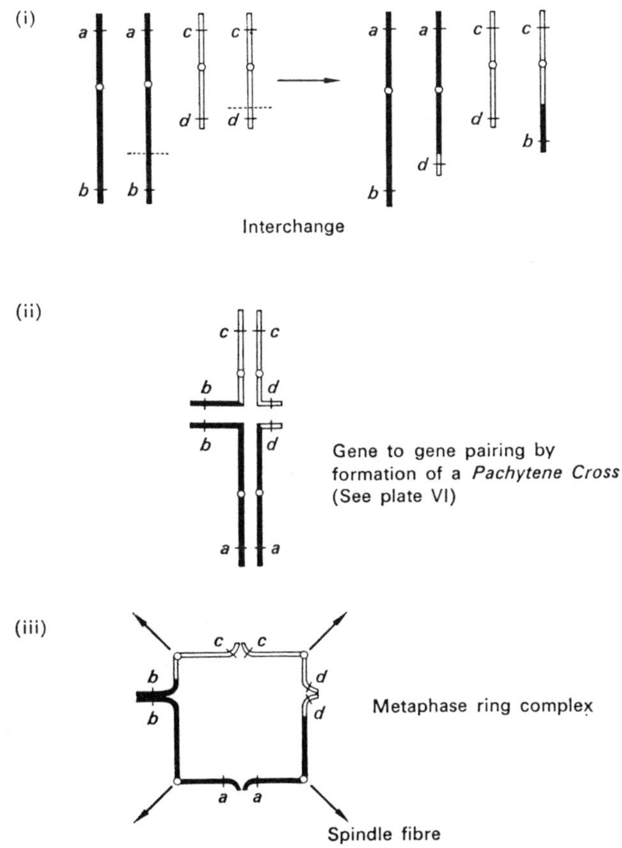

FIG. 67

The formation of a translocation, and subsequent formation, of Pachytene Cross and Metaphase ring complex

A remarkable example of this ring formation is found in the genus *Oenothera*, the evening primroses. These have a karyotype of fourteen chromosomes and in *Oenothera lamarckiana* twelve are involved in a ring formation, the other pair being separate.

As meiosis proceeds separation of the bivalents takes place, the consequence of which can be seen if we reconsider our original plant with four chromosomes (two pairs), involved in the ring (Figure 68). If adjacent chromosomes pass to the same poles the

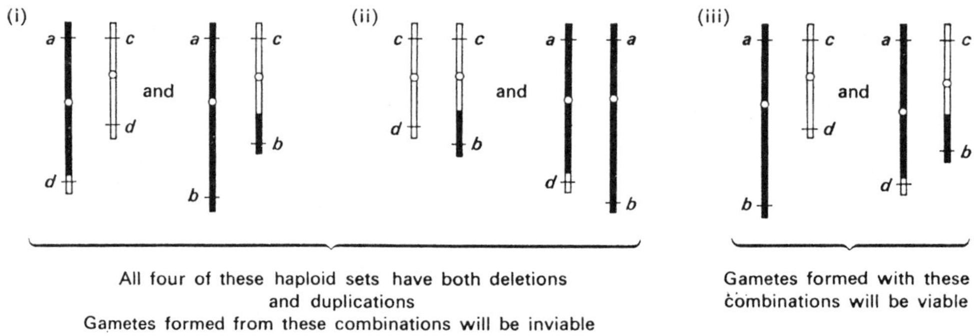

FIG. 68

The possible products from the ring complex shown in Fig. 67(iii)

gametes will contain some duplicated genes and some deletions, these gametes will be inviable. However if alternate chromosomes move to the same pole, two types of gamete are formed, each with a full complement of genes. These gametes will be viable. This means that one effect of translocation is partial sterility, which may be used as a means of detecting translocations.

(c) *Deficiencies.* This kind of chromosomal mutation involves the loss of a segment, with the subsequent rejoining of the remainder of the chromosome. If only a small segment is missing an organism in which only one of a pair of homologous chromosomes is so affected may be viable. The missing segment may carry dominant genes in which case any recessive alleles on the equivalent portion of the homologue can now express themselves. Homozygous deficiences are usually lethal.

(d) *Duplications.* A chromosome which has a segment missing, that is one which is carrying a deficiency, may instead of completing a set at one spindle pole, pass as an extra fragment to the other pole. This cell will thus be carrying duplicated genes. The extra fragment, if it has its own centromere, will remain separate, but if the centromere is missing it may become attached to another complete chromosome.

6.4.2. *Changes involving whole chromosomes*

Normally during anaphase I of meiosis the homologous chromosomes repel each other and become separated. If, however, instead of moving to opposite poles one pair move to the same pole, gametes would be formed which carried an extra chromosome. In the diploid organism formed from such a gamete that chromosome would be represented in triplicate. The organism would have a chromosome number of $2n+1$. Usually there is an effect on the phenotype. In *Drosophila*, for example, an extra fourth chromosome produces a fly which is small, weak, develops more slowly and has smaller, roughened eyes. Mongolism, or Down's syndrome, in human beings is also caused by the inclusion of an extra chromosome (Chapter 9). By the same means, that is the non-separation of homologues at meiosis (called *non-disjunction*), even more whole chromosomes can be present, extra to the normal set. In most cases this increases the deleterious effects.

6.4.3. *Changes involving whole sets of chromosomes* (*Polyploidy*)

If, during meiosis, the duplicated chromosomes fail to separate before the cytoplasm divides, and the new cell wall is formed, a cell will arise which contains twice the normal number of chromosomes, that is two sets instead of one. If this cell forms a gamete and ultimately fuses with a normal haploid gamete, the resulting offspring will have three sets of chromosomes instead of two, that is, it will be triploid not diploid. Fertilisation of a diploid gamete by another diploid would produce a tetraploid condition, and so on. This duplication of whole sets of chromosomes may also occur in the somatic tissue ($2n$) by the failure of a cell wall to separate the two identical groups of chromosomes formed during mitosis, resulting in a tetraploid somatic cell which could give rise to tetraploid tissue. Should this tetraploid tissue give rise to germ tissue, gametes will be formed with a diploid complement. It is probable that the latter is the more frequent source of polyploidy. Polyploidy is much more common in plants than animals (Plate VII).

(a) *Autopolyploids.* In this form only one species is involved and this means, in a tetraploid for example, a given chromosome will be represented four times. These individuals are often not drastically different from the parents except that they display increased vigour, are often larger and, in the case of plants, a darker green. Triploid *Drosophila* are larger and more robust than the diploids.

(b) *Allo-polyploids.* These are hybrid plants in which polyploidy has occurred. A hybrid is formed by crossing different species, sometimes even different genera. Each species will donate one set of chromosomes. Even in closely related species the sets of chromosomes, although similar to a certain degree, will not be identical. This means that at meiosis pairing cannot occur as no chromosome has a homologue present. Meiosis is blocked and the plant is usually infertile. If, however, mitosis occurs without cell division in the hybrid plant's cells, then the resulting cell would contain two identical sets of *both* of the original sets of chromosomes. In other words each of the chromosomes would now have a homologue. This is called the *allo-tetraploid* condition, and should these cells give rise to a germ tissue, gametes can be formed as meiosis can take place with the normal pairing of homologues. The gametes will contain one set of chromosomes from each parent that is, they will be *allo-diploid.* Fertilisation by another allo-diploid gamete will then reconstitute the allo-tetraploid condition. This means that allo-polyploid plants are usually fertile and true breeding, and are in effect a new species. An example of this is found in the *Primula* genus, where the hybrid formed between *P. floribunda* and *P. verticillata*, called *P. kewensis*, was for many years sterile. One of the inflorescences, however, set seed, in other words these flowers were fertile. Plants grown from the seed were larger and more vigorous than the hybrids and they were fertile. These plants were in fact tetraploids, with a complement of eighteen *pairs* of chromosomes instead of the infertile hybrid's complement of nine (floribunda) and nine (verticillata). Duplication of each set by a failure of the cell wall to form between mitotic groups had given rise to the tetraploid complement. As a source of new species, allo-polyploidy has played an important part in plant evolution, as well as in the development of new varieties of crop plants (Chapter 9).

In summary, gene mutation is the basic source of variation and in time a population will build up a fund of mutated genes, so that any individual will carry along its chromosomes a series of alleles most of which will be in the heterozygous state. Variations will also be produced by the chromosomal mutations we have described. Given this array of mutations new phenotypes can now be formed by reshuffling the genes in a number of ways.

6.5.1. *Independent segregation (assortment) of chromosomes*

This has been discussed in Chapter 4. During the first metaphase of meiosis the bivalents arrange themselves on the equator of the spindle. The spindle fibres contract and the homologous chromosomes are separated. It is a completely random process as to which pole of the spindle the homologue are drawn, the final outcome is two similar sets at both poles. However, the chance factor in the segregation to the poles means that a number of recombinations are possible. In the diploid cells in *Drosophila* there will be four pairs of chromosomes, one set (1♂ . . . 4♂) originally from the male parent, the other (1♀ . . . 4♀) from the female. During meiosis the homologues will separate out into the haploid sets. It is possible that the parental combinations will be reformed, but it is more likely that a range of combinations will result. In the case of *Drosophila* with a karyotype of four pairs sixteen combinations are possible (see Fig. 69). In humans with a karyotype of twenty-three pairs the number of possible combinations leaps to 2^{23}.

6.5.2. *Crossing over*

The number of variations formed by reassortment may be further increased if crossing over takes place during meiosis. If one chiasma forms in a bivalent, four chromosomes emerge, each differing not in the sequence of genes but in the overall combination of alleles (see Chapter 5). Exchange in this way of linked genes will result in a recombination

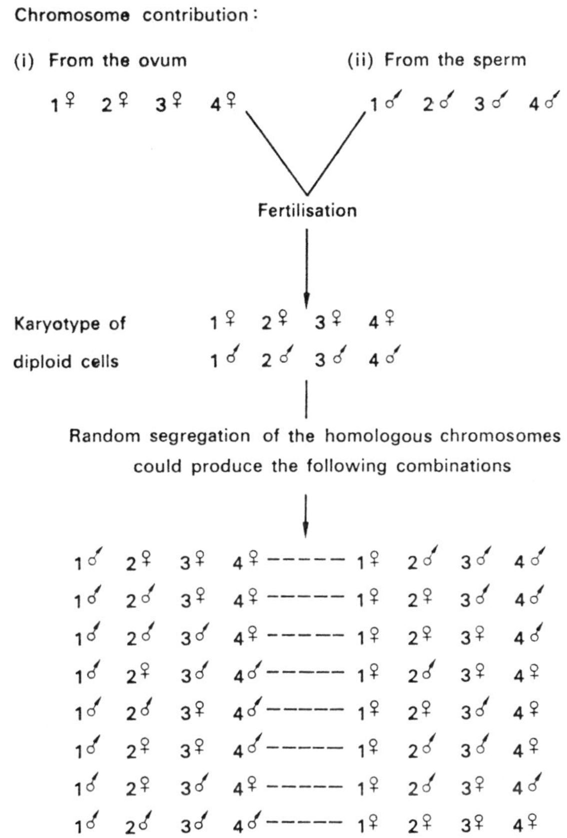

FIG. 69

To show the possible combinations of the parental chromosomes formed during gametogenesis in *Drosophila*

of genes between homologous chromosomes. As chiasma formation can occur at any point along the chromosome this in turn means that the number of different combinations of alleles on the chromosome is further increased (see Fig. 70).

6.5.3. *Random fusion of gametes*

The independent assortment of the chromosomes and the formation of alternatives by recombination and crossing over will all provide a vast array of haploid genotypes, in the form of gametes.

The genotype of the offspring, however, is dependent on the genetic material donated by both parents and as the fusion of gametes is a random process another source of variation is introduced. If each parent produced only two types of gamete random fertilisation would result in four sorts of offspring. *Drosophila* could form

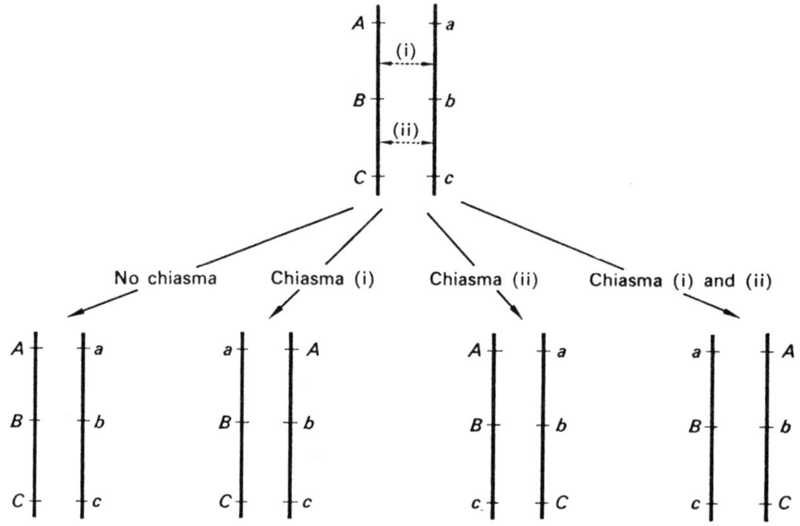

FIG. 70
Re-combination of alleles along a chromosome by crossing-over

sixteen types of gamete by independent assortment alone, random fertilisation could then theoretically result in 16^2, that is 256 zygotes.

6.6. MUTAGENS

In 1927 H. J. Muller discovered that the mutation rate in *Drosophila* could be greatly increased by exposing the flies to X-rays. This discovery was corroborated in 1928 by Stadler who used X-rays to induce the formation of mutants in barley. Besides X-rays, mutations can be induced by other ionising radiations, the α, β and γ rays. The effect of radiation is twofold. A direct hit on the chromosome may cause damage on the molecular level, that is damage to a gene. It may also fragment the chromosome producing the chromosomal mutations discussed previously in this chapter. There is also, however, a second indirect effect of radiation on the aqueous medium surrounding the chromosomes. If the ionising radiations hit a water molecule the energy of the radiation may displace an electron from the water molecule forming H and OH radicals.

$$H_2O \longrightarrow H_2\overset{+}{O} + e \text{ (electron)}$$
$$H_2O + e \longrightarrow H_2\overset{-}{O}$$
$$H_2\overset{+}{O} \longrightarrow H^+ + OH \text{ (radical)}$$
$$H_2\overset{-}{O} \longrightarrow OH^- + H \text{ (radical)}$$

These radicals in turn will give rise to HO_2 radicals which are highly reactive and are capable of damaging surrounding organic molecules, including the DNA molecules.

$$H + H_2O \longrightarrow H_2 + OH \text{ (radical)}$$
$$OH + OH \longrightarrow H_2O_2$$
$$H_2O_2 + OH \longrightarrow (HO_2) + H_2O$$

Damage to the DNA may occur to such an extent that not only gene mutation is likely but chromosome fragmentation may also take place.

Not all damage to the chromosomes will result in mutations, for the great majority of chromosome breaks are repaired in their original form, a very small percentage in fact form mutations.

Damage to the genetic material may also be brought about by chemicals such as mustard gas and hydrogen peroxide. The action of hydrogen peroxide may be blocked by the enzyme catalase present in many cells; catalase, therefore, is acting as an *anti-mutagen.*

An interesting mutagen is colchicine, a compound produced from the plant *Colchicum autumnale*. Mitotic cells treated with this chemical are anaesthetised in the metaphase condition. Separation of the two chromatids on each chromosome fails to occur and the cell formed will contain twice the number of chromosomes, that is it will be tetraploid. Colchicine and closely related drugs have been used to induce auto-polyploidy in a great many plants. The compounds are also used when details of the chromosomes of a given tissue are required. The dividing cells treated with colchinine will stop dividing and remain in the metaphase condition allowing the chromosomes to be examined. Plate IId is in fact a photograph of mitotic chromosomes which have been treated with colcimid, a related compound; it has disrupted the mitotic spindle and the cell is frozen in the metaphase condition. This technique is used extensively in hospitals when human chromosomes are examined for structural alterations which might lead to congenital deformity in the offspring (see Chapter 10).

7
Mating systems

7.1. INTRODUCTION

WE briefly considered the two main types of reproduction, sexual and asexual in Chapter 1. Asexual reproduction gives rise to offspring which are identical to each other and to the parent organism. This is because the basic mode of cell division involved is mitotic and because there is no gamete fusion during which genes may recombine. In sexual reproduction specialised reproductive cells fuse, and at some stage in the life history there is a meiotic division which gives rise to reassortment of genes. Both these factors lead to the formation of a variable progeny. Sexual reproduction itself, however, includes a wide variety of reproductive systems, each of which will have far-reaching effects upon the offspring,

The position of the meiotic division in the sexual cycle varies. In the vast majority of animals, and some algae and fungi, it takes place during the formation of gametes. In this case the major stage of the life history will be diploid, only the gametes will be haploid. In many algae (such as *Chlamydomonas, Volvox, Vaucheria*) and fungi (for instance *Sordaria* and *Cystopus*) and also in certain protozoa (such as *Monocystis*) meiosis takes place after the formation of the zygotic nucleus. In this case the zygote is the only diploid stage in the life history, the rest being haploid. In the majority of plants, including some algae and fungi, but mainly involving the Bryophytes, Pteriodophytes, Gymnosperms and Angiosperms, there are distinct haploid and diploid phases to the life history. The diploid phase produces haploid spores by meiosis, which germinate into the haploid phase. This then produces haploid gametes by mitosis, fusion of which restores the diploid condition. Where clear haploid and diploid phases are recognisable, the life history is said to show alternation of generations. In Gymnosperms and Angiosperms the haploid phase in the life history is very much reduced and the diploid phase is dominant.

In most organisms which reproduce sexually there is some degree of differentiation between the sexes. A few algae and fungi produce gametes which appear to be identical, that is, isogamous, but in fact in many of these physiological differences have been shown to exist between the gametes. In most cases there are clear differences between the gametes, some of which are described in Chapter 1. This condition is termed anisogamy, and in such circumstances fusion of two different types of gamete is required. A number of organisms are known where both male and female gametes are produced within the same body (hermaphrodite, or monoecious). In higher organisms, particularly animals, there has been a pronounced evolution towards the sexual differentiation of adult organisms so that each is responsible for the formation of one type of gamete. Examples are found, however, in lower organisms, such as *Mucor*, a heterothallic fungus which requires the presence of two separate mating strains to allow sexual reproduction to take place.

The main aim of this chapter is to use the above basic outlines of sexual reproduction to show how the degree of variability in the offspring can be affected by the precise mode of reproduction. We have seen in Chapter 6 that the basic sources of variability are found in the reassortment of genes during meiosis, re-combination during crossing over and the fusion of gametes carrying variable gene complements. Mutation is a largely random process and therefore, although a vital source of variation, lies outside the scope of this consideration.

It is clear that gametes produced by the same or closely related parents will have more similar genotypes than those produced by unrelated parents. This in turn will reduce the variability of the progeny of the former group compared with the latter group. The former system we term *inbreeding*, where progeny are formed by the fusion of gametes from the same or closely related parents. Where progeny are formed by the fusion of gametes from unrelated parents it is termed *outbreeding*. It must be clearly stated that these two terms are entirely relative. For example in human reproduction the offspring of first-cousin marriages are more inbred than those of second- or third-cousin marriages, but more outbred than the offspring of brother-sister marriages would be. This, in turn, would not be as inbred as the self-fertilisation of hermaphrodite plants or animals. Thus we can recognise one fixed point: it is that self-fertilisation represents the ultimate in inbreeding, and extension outwards leads to a gradual increase in the degree of outbreeding.

The main genetic consequence of outbreeding is that a population is formed in which the individuals show a pattern of variation with respect to a huge variety of characteristics. This reservoir of variability, some of it expressed and some of it hidden in the form of recessive alleles in heterozygous organisms, gives an outbreeding population a latent adaptability. If changes take place in the population's environment the chances are that some of the 'variants' within the population will be able to cope with the change and ensure the continuation of the population. On the other hand, an inbreeding population tends to give rise to groups of organisms with rather fixed genotypes. The reasons for this are explained later. Such populations will be well suited to the environment in which they find themselves, and will generally be able to make rapid and successful colonisation. If the environment changes however they may be unable to adapt to the changed situation as successfully as an outbreeding population since they have less genetic variability.

7.2. THE INCIDENCE OF INBREEDING

Extreme inbreeding as represented by self-fertilisation is relatively uncommon, particularly in the Animal Kingdom. Probably the only genuine examples of self-fertilisation in animals are the parasitic tapeworms. Care must be taken to distinguish self-fertilisation from parthenogenetic reproduction in for example Rotifers, Aphids, *Daphnia*, in which there is no fusion of gametes. In tapeworms eggs are fertilised by sperm from the same animal, although possibly from a different proglottid. Clearly this is an adaptation to its parasitic mode of life, in which the chances of cross-fertilisation with another tapeworm are almost negligible.

Perpetual self-fertilisation occurs in a number of plants. In flowering plants this may occur by the transference of pollen to the stigma of the same flower, or to the stigma of a separate flower on the same plant. In some cases self-fertilisation takes place even before the flowers open; a phenomenon termed *cleistogamy*, such as in the wood sorrel. Other examples of perpetual self-fertilisers are oats, peas, beans, tomatoes and quite a large number of species of grasses. Stebbins has classified a number of species of grass according to their pattern of fertilisation and life history:

TABLE 4. (modified from Stebbins *Variation and Evolution in Plants*)

	Perennials	*Annuals*
Almost always self-fertilisation	5	37
Some self-fertilisation Some cross-fertilisation	40	3
Always cross-fertilisation	26	0

There appears to be a tendency for the annuals to self-fertilise and the perennials to cross-fertilise some or all of the time. It is suggested that annuals cannot afford to risk cross-fertilisation, since if they fail to set seed they will perish. Perennials live for several years and can therefore afford the more risky process of cross-fertilisation in order to obtain variability. Annuals tend to sacrifice variability in order to ensure the production of seed.

7.3. DELETERIOUS EFFECTS OF INBREEDING

Plant species such as maize and rye, which are typical cross-fertilisers, can be induced to self-fertilise. The first effect of such action is to drastically reduce the amount of seed set. Plants which are raised from this seed will be significantly less vigorous in growth than the parental stock. If inbreeding is continued for successive generations this decrease in vigour becomes gradually more pronounced until what is termed the *inbreeding minimum* is reached. After this point, continued inbreeding does not lead to any further degeneration. Before this point is reached, however, the original species will have separated into a number of more or less distinct lines, each with characteristic properties. Each line will show a decrease in vigour when compared with the parental stock, although the extent of the decrease will vary.

In the vast majority of animals the possibility of self-fertilisation has been eliminated, but similar effects are obtained by matings between brothers and sisters, so called *sib-mating*. Sib-mating in experimental animals such as rats and *Drosophila* leads to an inbreeding minimum of the same kind as found in plants. The number of generations required to reach the inbreeding minimum is, however, much greater in sib-matings than in self-fertilisation. Also in animals the characteristic decrease in vigour may not be found in all the lines produced by successive inbreeding; some may have normal vitality.

In man the situation is more or less the same, although the degree of inbreeding has been greatly reduced by religious and cultural barriers in most societies. For instance, sib-mating is not tolerated in most societies, and in certain cases this prohibition is extended to first-cousin marriages as in some States of America. Such prohibitions have not always been the case. The Ptolemy Dynasty of Ancient Egypt was maintained by sib-mating to prevent dilution of the 'royal blood'. A certain degree of inbreeding is characteristic of small populations which are isolated either geographically, as on islands or valleys, or socially, as by nationality or religion. In a study of the populations of two islands off the west coast of Sweden, it was found that a particular hereditary mental disease occurred at high frequency in these populations. In fact the frequency of many hereditary defects or diseases seems to be higher among isolated populations than it is in the main population. This indicates that a high frequency of deleterious effects may be associated with inbreeding.

It also seems likely that inbreeding in man, as in other animals and plants, leads to the formation of distinct lines, which may show considerable differences from one another. This can be seen in the South American Andes, where there are large numbers of Indian tribes living in the different valleys and regions. Substantial differences in physical features, mental abilities and general viability are said to occur between the tribes. It is suggested that the existence of these tribes may be due to inbreeding within small isolated populations, giving a number of distinct genetic lines.

7.4. INBREEDING LEADING TO HOMOZYGOSIS

We can conclude from the previous section that inbreeding leads to an increase in the appearance of deleterious traits. We must now determine whether it is inbreeding itself which is harmful, or whether it leads to further effects which are responsible for the appearance of degenerative characteristics.

The fact that varying degrees of inbreeding are used either perpetually or occasionally by a significant number of organisms would suggest that, in itself, it is not harmful. We have already come across several groups of plants where self-fertilisation is the rule, and has no obvious harmful effects. In the breeding of domestic animals, particularly dogs, a high degree of inbreeding is used and this does not always have adverse consequences. Although the example of domesticated animals is quoted here, it must be used with care. The characters for which these animals are bred may be attractive or useful as far as man is concerned, but many of them would certainly be harmful if the animals in question were taken out of their highly protected 'domestic' environment.

The relationship between inbreeding and the appearance of deleterious characters lies in the fact that it tends to increase the proportion of homozygous organisms within a population (that is its *homozygosity*). Consequently it will also reduce the proportion of heterozygous organisms (that is the *heterozygosity* of the population). Figure 71 shows a theoretical series of crosses involving the alleles *A* and *a*. Starting with an entirely heterozygous population and carrying out successive 'selfings', we find that with each subsequent generation the proportion of heterozygotes falls, and the proportion of homozygotes rises. Here we are considering only one pair of alleles, whereas all organisms will contain a huge number of pairs of alleles and will be heterozygous for a large number of them. Successive inbreeding leads therefore to the production of organisms which tend to be homozygous for many of their allelic pairs, possessing either the double dominant or the double recessive alleles.

All organisms possess alleles which produce harmful phenotypic characters. The effect will vary from cases where the allele causes a marginal phenotypic effect to those which are lethal. If the allele concerned is dominant then it will be expressed in the phenotype of the heterozygous organism and will be removed from the population at a rate depending on the severity of the effect. If the deleterious allele is recessive, however, it will not appear in the phenotype of the heterozygote, only in the homozygous recessive. We have seen that inbreeding leads to the formation of organisms which are homozygous for particular alleles and therefore it is much more likely that recessive, deleterious alleles will be expressed in the phenotype of an inbred population. It should be made clear that all recessives are free to appear in this way, and occasionally beneficial ones come to light. Also inbreeding does not create harmful recessive alleles, it merely leads to increased homozygosity which allows them to show up in the phenotype.

7.5. THE FIXATION OF CHARACTERS

The fact that inbreeding leads to increased homozygosis can also be used to explain the appearance of several distinct genetic lines among inbred populations. If we take an

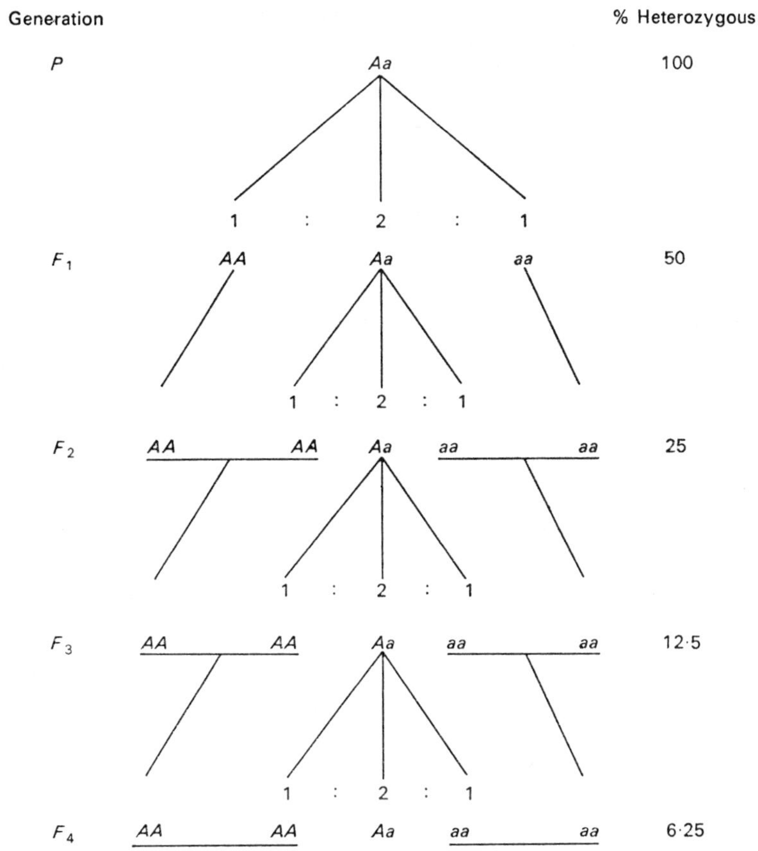

FIG. 71

The effect of successive inbreeding on the proportion of heterozygotes to homozygotes in a population

organism which is heterozygous for two pairs of alleles *AaBb*, successive inbreeding will eventually result in four homozygous lines, *AABB*, *aaBB*, *AAbb*, *aabb* being formed. These lines would be phenotypically distinct and would breed true within themselves. Thus from our originally heterozygous type, inbreeding has led to a fixation of characters to form four pure lines. Clearly many more than two pairs of alleles will be involved in any practical example, but the principle can still be applied. Of course, the larger the number of allelic pairs, the larger the number of separate lines which can theoretically be produced by inbreeding.

Such a fixation of genetic characters in distinct lines is of great importance from an evolutionary point of view. The formation of a new species is the key step in evolution, as is explained in Chapter 8. In order for speciation to take place some form of isolation of one breeding group from another such group is required. This prevents the exchange of genetic information between the two groups, and allows them to become adapted to the environment along different lines.

While it would clearly be a mistake to refer to the distinct lines produced as a result of inbreeding as separate species, it can be thought of as one mechanism by which reproductive isolation may be created. Clearly in those plants which perpetually self-fertilise the concept of a reproductively isolated group which may go on to form a species is not a valid one. However, in organisms which may inbreed occasionally, perhaps as a response to physical isolation or environmental stress, the formation of distinct lines may provide a basis upon which further differentiation into species may occur.

7.6. THE INCIDENCE OF OUTBREEDING

Apart from the very rare exceptions mentioned earlier, all animals are outbreeding to some degree. This is not only true of those which show clear sexual differentiation into male and female organisms, but also in hermaphrodites such as *Hydra*, *Lumbricus*, *Helix* and so on. Many of the fungi and algae possess distinct mating strains analogous to the separate sexes of higher organisms. In such heterothallic organisms sexual reproduction will only take place between the two mating strains thus ensuring outbreeding. In the majority of higher plants outbreeding is the most common mode of reproduction and often there are elaborate mechanisms to ensure its occurrence.

7.7. MECHANISMS ENSURING OUTBREEDING

'Nature abhors perpetual self-fertilisation' said Charles Darwin, who was evidently impressed by the variety and complexity of the mechanisms by which organisms ensure outbreeding.

(a) *Dioecism* describes the condition where the male and female gametes are produced on separate individuals. Where this is the case self-fertilisation is impossible. The majority of animals show dioecism and in the higher plants this has developed into a higher degree of sexual differentiation between the male and female organisms. Such organisms have complex genetic mechanisms controlling sex which ensure that male and female offspring are produced in equal numbers (Chapter 5). The production of gametes in completely separate organisms means that there is an element of risk in the processes of fertilisation and formation of progeny. The mobility of animals and the development of complex control systems have enabled them to largely overcome this problem by the evolution of suitable mating behaviour. In most cases such behaviour ensures the fusion of the gametes. The risk in higher plants, however, is much greater, since they must rely on external agencies, such as wind, insects, and so on, to transfer the male gametes to the female structures. For this reason we find that the number of dioecious plants is small when compared with those with either hermaphrodite flowers, or both male and female flowers on the same plant.

(b) *Protandry and protogyny*. In hermaphrodite organisms self-fertilisation is often precluded by the maturation of the male and female gametes at different times. In plants such as the Compositae and Labiatae and in some species of the coelenterate animal *Hydra*, the male gametes reach maturity and are released long before the female gametes of the same organism are developed, Such organisms are said to be *protandrous*. More rarely one finds plants such as *Luzula* (woodrush) and *Scrophularia* (figwort) where the female gametes are fully developed before the anthers are mature. Such flowers are termed *protogynous*. Both these conditions will tend to ensure outbreeding, but in the case of monoecious plants, with male and female flowers on the same plant, it is perfectly possible to have flowers at different stages of development on the same plant. In this case, of course, self-fertilisation can easily take place.

(c) *Self-incompatibility*. In many plants there are genetically controlled systems which ensure that pollen will only develop on the stigma of another plant. If the pollen is placed on the stigma of the same flower it fails to develop sufficiently to bring about fertilisation. The classic example of such self-incompatibility is found in species of *Primula*. In *Primula* two distinct types of flower are found, the pin and thrum forms. This condition is termed *heterostyly*. In the pin form the style is long and the anthers are attached halfway down the corolla tube. In the thrum form the style is short and the anthers are attached at the mouth of the corolla tube (Fig. 72). The occurrence of these two types of flower structure forms the basis of an outbreeding mechanism. The position of the anthers and the length of the style are controlled by two genes which are linked to a gene which controls the growth of the pollen tube. The thrum type is heterozygous

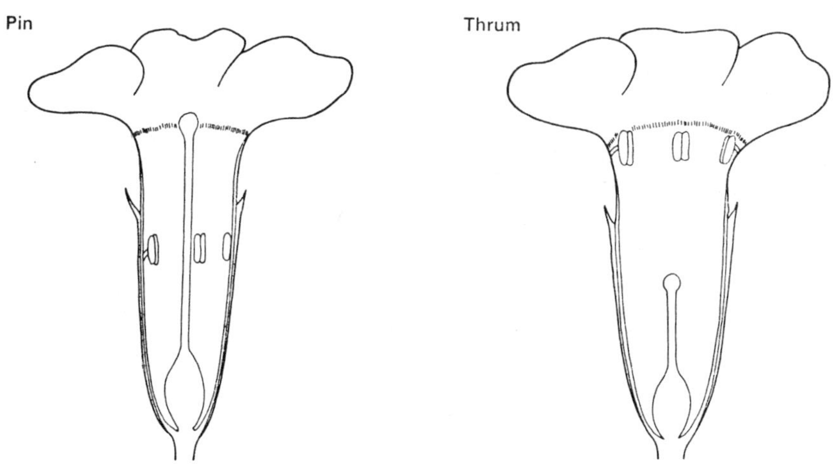

FIG. 72
Primula. Vertical sections of the pin and thrum forms of flower

for all three genes, whereas the pin type is homozygous. Considering simply the gene controlling pollen tube growth, the thrum type is therefore *Ss*, and the pin type *ss*. A cross between the two types of flower gives equal numbers of pin and thrum offspring. The two types of pollen produced by the thrum flower, *S* and *s*, are both compatible with the style of the pin flower whose genotype is *ss*. Similarly the pollen produced by the pin flower, all *s*, is compatible with the style of the thrum flower whose genotype is *Ss*. However, both the *S* and *s* pollen grains produced by the thrum flower are incompatible with the *Ss* style of the thrum flower. Similarly the *s* pollen of the pin flower is incompatible with the *ss* style of the pin flower, although to a noticeably lesser degree. Thus there is in operation a mechanism which tends to ensure crosses between pin and thrum flowers only, and prevents *thrum* × *thrum* and *pin* × *pin* crosses. It also appears to be the characteristics of the style, inhibiting the proper growth of the pollen tube, which is the cause of the self-incompatibility in this case.

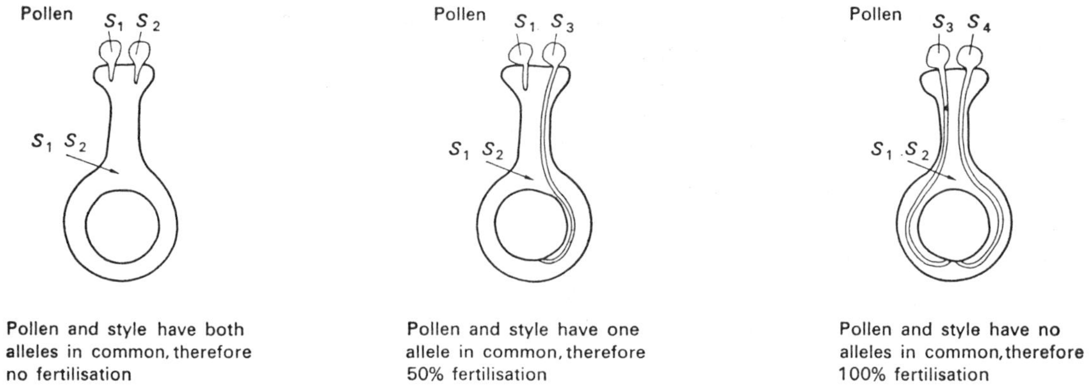

FIG. 73
Self-sterility controlled by multiple alleles

In other cases of self-incompatibility it appears that it is the genotype of the pollen grain which is of importance. In plants such as *Petunia*, *Trifolium* (clover) and *Oenothera* (Evening primroses) the locus controlling pollen tube growth *S*, has many alleles, *S1*, *S2*, *S3*, *S4*, *S5* and so on. In red clover forty such alleles are known. A single individual will contain only two of these alleles, and will, therefore, produce one or two types of pollen. If the pollen grain and the style have no alleles in common, then 100% growth of pollen tubes takes place. If the pollen grain and the style carry alleles in common, then growth does not take place, Fig. 73. This type of self-sterility is common among cultivated crops such as cherries and apples. In these cases it is essential that compatible varieties are grown in close proximity to ensure fertilisation and a good crop (Plate VIII).

7.8. HYBRID VIGOUR (HETEROSIS)

We have seen that inbreeding in most plants and animals causes a marked reduction in vigour in the offspring. Such a reduction may have many phenotypic manifestations. In complete contrast, when inbred lines are 'outcrossed' the *F1* hybrids show a tremendous increase in vigour and viability. Even if the inbred lines have, after many years of inbreeding, become extremely degenerate the *F1* hybrids will at once show a return to full vigour. This phenomenon is termed *hybrid vigour*, or *heterosis*.

Heterosis is of great practical importance. *F1* hybrids are widely used in a large variety of crop plants, where their increased vigour leads to greater yields. The yields of maize varieties, grown as a staple food crop in many parts of the world, have been multiplied by the use of *F1* hybrids produced by outcrossing inbred lines. Plants such as tomatoes, which are spontaneously self-fertilising, can be artificially outcrossed to produce heavy-yielding *F1* hybrids. Heterosis has also been observed in animals, indeed a possible example has been cited in man. A. Dahlberg has pointed out that the increase in the average height of the Swedish population during the last century may be due to heterosis. The isolation of communities has tended to break down with improved communications thus leading to an increase in the degree of outbreeding. The outbreeding of previously inbred lines has led to heterosis expressed as an increase in height. Care must be taken in adopting this interpretation, since it is probable that environmental factors such as nutrition have also altered during this period, but it is an interesting suggestion.

It should not be assumed that heterosis is manifested only as an increase in size or vigour of growth. In soybean hybrids an increase in the numbers of pods and seeds is seen, but there is no difference in overall size. In poultry, crosses between White Leghorn and Barred Plymouth Rock strains show an increased fertility of the eggs produced.

There have been a number of theories explaining heterosis. In 1908 Shull and East proposed that the heterozygotic state is 'stimulating' when compared with the homozygotic state. In other words, with respect to alleles A and a, Aa will, in some way, be superior to either AA or aa. The theory does little more than restate the facts of heterosis, since inbred lines will show a high degree of homozygosity and crosses between them are bound to be heterozygous at most loci. The most likely theory to emerge in explanation of heterosis involves the interaction of different dominant alleles. It has been suggested that the loss of vigour on inbreeding is due to the emergence of organisms carrying harmful recessive alleles in the homozygous state. Therefore at many loci the dominant allele will be absent. It has been suggested that a number of these dominant alleles control the general vigour of the organism, and that the absence of some of them reduces vigour. Clearly the same combinations of dominant alleles will not be missing in all the offspring, accounting for the formation of different lines and the variation in the degree of degeneration. If we consider a cross between two such inbred lines, having five loci the genes at these loci have an effect on vigour. (The number is in practical circumstances much larger.)

Inbred Line I		Inbred Line II
AABBccddEE	×	*aabbCCDDee*
	↓	
F1 hybrid	*AaBbCcDdEe*	

Thus the hybrids will contain a much greater number of the favourable dominant alleles controlling the vigour of the organism. Interaction of these dominant alleles will cause heterosis.

The main objection to this theory is that on inbreeding of the above hybrids one may expect to obtain organisms which are homozygous dominant, that is *AABBCC-DDEE*. These organisms would be expected to have at least as good vigour as the hybrids, whereas in fact, in plants at least, all inbred lines show a decline in vigour. However, if one considers the number of 'vigour genes' which are likely to be involved, the mathematical probability of a line forming which is homozygous for all the dominant alleles is minute. Also some of the vigour alleles are likely to be linked on the same chromosome, preventing the independent segregation necessary to form organisms homozygous for all the vigour genes.

8

Genes in populations and selection

8.1 INTRODUCTION

IN our consideration of inheritance so far we have largely been concerned with the behaviour of genes in individuals, or small groups or organisms. A geneticist investigating a particular gene will make crosses involving organisms with known, or suspected, genotypes. He will look for certain phenotypic ratios in the offspring predicted by the laws of inheritance. Such ratios may be interpreted as information about the gene under study. The whole situation, the alleles present, their frequency, and the mating system adopted is rigidly controlled. This must, of course, be the case if reliable results are to be obtained. We must not, however, assume that such carefully controlled conditions exist in natural groups of organisms.

Before going on to look at some of the ways in which artificial and natural groups of organisms differ, the idea of a *population* must be introduced. In experimental work the population of organisms is well defined, the *Drosophila* in a milk bottle, the locusts in a cage, the pea plants in a garden. The population is thus a group of organisms of the same species which are freely interbreeding, and which are not interbreeding with other such groups. Although the distinction between such groups is not as clear in natural circumstances, the definition of a population holds good. For example in man an isolated tribe of natives in New Guinea constitutes a population, as does a group of Bedouin Arabs. The two groups are of the same species but do not freely interbreed with each other. In many animals, particularly those with limited powers of movement, and in plants the division of a species into populations is on a much more localised basis. In the snail *Cepaea* it has been estimated that specimens collected 100 yards apart can effectively be considered as members of separate populations. In flowering plants the limits of a population will largely be set by the efficiency of the pollination mechanism. It must be made clear that the division of a species into populations does not imply a reproductive barrier. If two separate populations are brought together then they will interbreed and produce offspring. A population is not, therefore, a taxonomic division, but one largely imposed by environmental factors such as climate, and geographical features such as seas, rivers, mountains and so on, or by the organism's powers of dispersal.

In artificial laboratory populations the frequency of alleles is carefully determined. In a cross between organisms *AA* and *aa* a population is created in which the frequency of allele *A* is 0·5, and of allele *a* 0·5. In the *F2* such a cross will give a genotypic ratio of 1*AA*:2*Aa*:1*aa*, with a corresponding 3:1 phenotypic ratio if *A* is dominant. In wild populations there is no guarantee that the alleles will be present in equal frequencies and this will mean that the phenotypic ratios, with respect to *A* and *a*, will not necessarily be 3:1. Also in such a laboratory cross the formation of a heterozygous *F1* and the 3:1 ratio in the *F2* will be ensured by crossing, for example, male *AA* with female *aa*. All matings must, therefore, be *AA* × *aa*. Clearly in wild populations no such 'mating compulsion' exists, and matings *AA* × *AA* and *aa* × *aa* are, in most cases, just as likely to occur. Again this will mean that, in the case of *A* and *a*, a phenotypic ratio of 3:1 is unlikely to be observed. This argument does not mean that the classic laws of inheritance are false, merely that we must apply them to a differing set of circumstances if we are to interpret the behaviour of genes in wild populations.

8.2. GENE FREQUENCIES

The appearance of particular phenotypic ratios in wild populations depends primarily on the frequency with which the genes concerned occur in the population. If we take a gene with two alleles, *A* and *a*, the frequency of the gene *as a whole* in the population is considered as unity, therefore the frequencies of the two alleles must add up to one.

$$\text{Let the frequency of } A = \mathrm{p}$$
$$\therefore \text{ let the frequency of } a = 1 - \mathrm{p}$$

Consider a cross between heterozygous organisms, *Aa*:

Female gametes

		p*A*	(1 − p)*a*
Male gametes	p*A*	p^2*AA*	p(1 − p)*Aa*
	(1 − p)*a*	p(1 − p)*Aa*	$(1-\mathrm{p})^2$*aa*

Thus the proportions of genotypes present will be:

AA	*Aa*	*aa*
p^2	$2\mathrm{p}(1-\mathrm{p})$	$(1-\mathrm{p})^2$

These frequencies will occur for a single pair of autosomal alleles, and will be maintained so long as random mating occurs. That the equilibrium will be maintained can be demonstrated by drawing a table of the genes contributed to a population by different genotypes:

		Gene frequency contributed	
		A	*a*
Genotype	*AA*	p^2	—
	Aa	$p(1-p)$	$p(1-p)$
	aa	—	$(1-p)^2$

Total frequency of $A = p^2 + p(1-p) = p^2 + p - p^2 = p$
Total frequency of $a = p(1-p) + (1-p)^2 = p - p^2 + 1 + p^2 - 2p = 1 - p$

Thus further random mating will only perpetuate the same proportion of alleles in the population. This equilibrium is termed the *Hardy-Weinberg equilibrium.* The provisor must be made that the gene frequency must not be altered by any other factor, such as mutation. If the allele *A* mutates to *a* at a given rate, then clearly the frequency of the alleles alters, as do the ratios of phenotypes. The immigration into the population of organisms carrying predominantly one of the two alleles will also serve to alter the overall frequencies. More will be said of factors altering gene frequencies later in this chapter.

Let us now examine a natural population and see whether observed frequencies agree with this distribution. In human red blood cells there are a large number of antigens, the presence of which are genetically controlled. The presence or absence of the antigens enables blood to be 'grouped' (Chapter 9). Two of the less well-known antigens which may or may not be present are the antigens M and N. To detect the presence of the antigens, use is made of antiserum produced in a rabbit. The antigen, for e.g. M, is injected into a rabbit whose blood then produces anti-M which will be present in the blood serum. If a small quantity of such serum is mixed with human blood in which the red cells contain antigen M, the blood agglutinates. A similar procedure is used to detect antigen N. In human beings the following phenotypes, and corresponding genotypes, are possible:

Antigens present	*genotype*
M	*MM*
N	*NN*
MN	*MN*

Thus the system is controlled by a single pair of alleles, and, since the heterozygote contains both antigens, there is no dominance. It has been found that the proportions of the phenotypes in a population varies in different regions of the world:

For example:

	M	*MN*	*N*	*Total tested*
American Pueblo Indians	83	46	11	140
Australian aborigines	3	44	55	102

(Boyd, *Tabulae Biologicae* 17:230, 235, 1939)

Clearly there are great differences in the frequency of the alleles concerned between the two populations, but do these frequencies agree with the Hardy-Weinberg equilibrium?

In the case of the Pueblo Indians:

Observed phenotypes		*genotypes*	*frequency of alleles* *M*	*N*
M	83	*MM*	166	—
MN	46	*MN*	46	46
N	11	*NN*	—	22
Total gene frequencies in population:			212	68
		=	0·76=p	0·24=1–p

From these values it is possible to calculate the proportion of phenotypes one would expect in a population of 140 according to the Hardy-Weinberg equilibrium.

$$\text{Frequency of genotype } MM = p^2 = 0{\cdot}76^2 = 0{\cdot}58$$
$$0{\cdot}58 \times 140 = 81$$
$$\text{Frequency of genotype } MN = 2p(1-p) = 1{\cdot}52 \times 0{\cdot}24 = 0{\cdot}36$$
$$0{\cdot}36 \times 140 = 51$$
$$\text{Frequency of genotype } NN = (1-p)^2 = 1 - 1{\cdot}52 + 0{\cdot}58 = 0{\cdot}06$$
$$0{\cdot}06 \times 140 = 8$$

Arranging the two sets of results together we see a close correspondence:

	MM	*MN*	*NN*
Observed numbers	83	46	11
Expected numbers	81	51	8

Using the identical calculation the expected frequency of the three groups in the Aboriginal population can also be calculated:

	MM	*MN*	*NN*
Observed numbers	3	44	55
Expected numbers	6	38	58

Thus both populations conform to the Hardy-Weinberg equilibrium even though the frequency of the alleles in the two populations is almost exactly reversed. The great importance of the Hardy-Weinberg equilibrium is that it allows us to show that a character in a natural population is genetically controlled, and is obeying the basic laws of inheritance. By taking into account the frequency of a particular gene in a population we can see that divergence from the classic ratios of Mendelism does not mean that the laws are not in operation.

In Chapter 10 a simple class exercise is suggested to demonstrate the Hardy-Weinberg equilibrium. People vary in their ability to taste the substance phenyl-thio-carbamide (P.T.C.). To some it tastes extremely bitter, while to others it is tasteless. The proportions of tasters to non-tasters varies between populations according to the frequencies of the allele *T* (for tasting) and *t* (for non-tasting). Although the situation is different from the MN blood antigens in that *T* is dominant to *t*, the principles are identical. Using the Hardy-Weinberg equilibrium it is possible to calculate the proportions of tasters which are homozygous and heterozygous.

Let us take a hypothetical population of 100, in which seventy are tasters (either *TT* or *Tt*) and thirty non-tasters (*tt*).

$$\begin{aligned} \text{The frequency of } tt &= (1-\text{p})^2 \\ 0{\cdot}3 &= (1-\text{p})^2 \\ \therefore \sqrt{0{\cdot}3} &= 1-\text{p} \\ 0{\cdot}55 &= 1-\text{p} \\ \therefore \text{p} &= 0{\cdot}45 \end{aligned}$$

$$\begin{aligned} \text{Since the frequency } TT &= \text{p}^2 \\ &= 0{\cdot}45^2 \\ &= 0{\cdot}2 \end{aligned}$$

$\therefore$ of the tasters 20 will be *TT*
and 50 will be *Tt*

This example is valuable in that it indicates that in cases where dominance of one allele over another occurs, a large proportion of the phenotypically dominant organisms are likely to be heterozygous. In the condition *albinism* in human beings the presence of a homozygous recessive allele causes an inability to produce the pigment melanin. The frequency of albinos is estimated at about 1 in 20 000.

$$\therefore\ (1-p)^2=\frac{1}{20\,000}$$

$$1-p=\frac{1}{141}$$

$$p=\frac{140}{141}$$

The expected frequency of heterozygotes $=2p\ (1-p)$

$$=2\times\frac{140}{141}\left(1-\frac{140}{141}\right)$$

$$=\text{approx}\ \frac{1}{70}$$

Thus the frequency of heterozygotes is 1 in 70, compared with 1 in 20 000 for the homozygous recessives, that is, for every albino in a population there will be $\frac{20\,000}{70}=286$ heterozygotes. In general it can be said that the rarer recessive allele, the greater the proportion of heterozygotes over homozygotes showing the allele (Table 5).

TABLE 5

I *Frequency of affected persons*	*II* *Frequency of carriers*	*Ratio of II:I*
1 in 10	1 in 2·3	4·3:1
1 in 100	1 in 5·6	18:1
1 in 1000	1 in 16	61:1
1 in 10 000	1 in 51	198:1
1 in 100 000	1 in 159	630:1
1 in 1 000 000	1 in 501	1998:1

Thus even recessive alleles which have lethal phenotypic effects can exist at quite a high frequency within a population in the heterozygous state.

8.3. MECHANISMS CAUSING CHANGES IN GENE FREQUENCY

The overall characteristics of a population of organisms are determined by the frequency of the various genes within the population. Any change in these characteristics will be brought about by a change in the gene frequencies. We have seen that, by application of the Hardy-Weinberg equilibrium, and assuming random mating, gene frequencies will remain the same from generation to generation. We can recognise, however, several agencies which can act upon this equilibrium and bring about a change in gene frequency.

8.3.1. *Mutation*

In Chapter 6 mutation has been described as a process by which the genetic material is altered. In its most basic terms a mutation is a change in the sequence of organic bases in the DNA constituting a particular gene. The effect of such a change varies, but in some cases it gives rise to an alternative form of the gene, an allele. The allele will give rise to a phenotypic character which differs from the 'normal' type.

Let us consider a population in which a gene exists only as the form *A*. The frequency of this gene will therefore be 1. If mutation takes place to give the allele *a*, then the frequency of *A* will fall below 1. If such a mutation was a completely isolated event, then the overall effect upon the gene frequency in the population would be negligible. There is, however, considerable evidence that mutation is a recurrent process, and that a particular gene mutates at a constant rate. This will lead to a gradually increasing frequency of *a* in the population and associated phenotypic changes. In fact the mutant allele *a* will eventually replace the original *A*. In most cases, however, the mutation is a reversible process, and the allele *a* will mutate back to *A*, although not necessarily at the same rate as *A* mutates to *a*. Eventually an equilibrium is established at which the frequency of the alleles *A* and *a* remains constant. The exact position of the equilibrium depends upon the rates at which *A* mutates to *a*, and *a* mutates to *A*.

The frequency of $A = \mathrm{p}$

The frequency of $a = 1 - \mathrm{p}$

Let the relative mutation rates be:

$$A \underset{3}{\overset{4}{\rightleftarrows}} a$$

Thus new *a* alleles are formed at the rate of 4p

New *A* alleles are formed at the rate of 3(1 − p)

Equilibrium is reached when these two rates of formation of new alleles are equal:

$$4\mathrm{p} = 3(1-\mathrm{p})$$

$$\therefore \mathrm{p} = \frac{3}{7} = \text{frequency of } A$$

$$1-\mathrm{p} = \frac{4}{7} = \text{frequency of } a$$

8.3.2. *Migration*

Clearly the frequency of a gene within a population will only remain constant while that population remains isolated from other populations. If the gene is continually flowing into and out of a population then its frequency is bound to fluctuate. In practical terms this process of *gene flow* can be thought of as the inclusion of new individuals into the breeding group, or the exclusion of some individuals from it. Immigration and emigration of organisms into and out of a population are clearly going to be of the greatest importance in those organisms which are highly mobile, that is, the majority of animals. In sedentary animals and plants it can occur by the dispersal of reproductive structures, although to a much lesser extent.

Let us consider a single case of the effect of migration on gene, genotype and phenotype frequencies. Consider an isolated population of eighty organisms all of which are homozygous for the recessive allele *a*. The gene frequency of *a* will therefore be 1, the genotype frequency of *aa* will be 1, and all the organisms will show the recessive character in their phenotype.

Suppose that twenty 'immigrant' organisms appear all of which are homozygous for the dominant allele *A*. In the total population of 100, the gene frequency of *A* is therefore 0·2, and of *a* 0·8. Assuming that random mating occurs between the 'resident' and 'immigrant' groups:

$$\text{frequency of } AA = p^2 = 0{\cdot}2^2 = 0{\cdot}04$$
$$\text{frequency of } Aa = 2p(1-p) = 0{\cdot}4 \times 0{\cdot}8 = 0{\cdot}32$$
$$\text{frequency of } aa = (1-p)^2 = 0{\cdot}8^2 = 0{\cdot}64$$

Thus equilibrium is established when 36% of the population are either *AA* or *Aa* and therefore exhibit the dominant allele *A* in their phenotype, and 64% are *aa* and exhibit the recessive allele. If the situation was reversed in that the resident population carried the dominant allele and the immigrants the recessive allele the effect on the phenotype of the population is far less spectacular. Using a similar calculation it can be shown that equilibrium is reached when 96% of the population are either *AA* or *Aa* and only 4% are *aa* and therefore exhibit the recessive allele phenotypically.

8.3.3. *Random genetic drift*

A number of geneticists, chiefly Sewall Wright, have argued that changes in the frequency of a gene in a population may be brought about by processes of chance alone.

In a classic experiment Wright set up artificial populations of *Drosophila* containing equal frequencies of the gene causing forked bristles and its wild-type allele. The gene for forked bristle, *f*, is recessive and sex-linked, that is it is carried on the *X* chromosome. The populations were set up as follows:

4 males	*4 females*
2 of *f Y*	1 of *f f*
2 of + *Y*	2 of *f* +
	1 of + +

Thus the frequency of both alleles is 0·5. He set up a total of 108 identical populations and at the end of each generation (every fourteen days) he selected, at random, four males and four females from each population to start a new culture. After sixteen generations the proportions of *f* and + genes in the populations were:

41 populations	All +	No *f*
29 populations	No +	All *f*
38 populations	Mixed + and *f* (not necessarily in equal proportions)	

Thus either gene could be completely eliminated from a population. This would indicate that there is no significant difference in the viability of flies carrying the different alleles and that the elimination is entirely due to chance. It must be made clear that this interpretation has been questioned on the grounds that minute differences in the advantage conferred by a particular allele are very difficult to detect and also that genes very often have a range of phenotypic effects. It is suggested that, for example in this case, the allele *f* may have physiological effects besides causing forked bristles. However it does seem likely that this process of random drift of genes may have some importance in altering gene frequency. It will give rise to changes in the frequency of a gene within a population which are due entirely to chance and are totally independent of changes in the frequency of the same gene in another population.

The key to the importance of random genetic drift lies in the size of the population. It seems likely that such drift caused purely by chance can only eliminate an allele completely from a relatively small population. The larger the population becomes the more marginal are the effects of random genetic drift.

Let us consider a population of 100 organisms in which an allele exists at a frequency of 0·01. In effect this means that there are two such alleles in the population, almost certainly in the heterozygous state. It is possible that these organisms fail to reproduce or the allele becomes lost in female polar bodies or unused sperm. The very low numbers of this allele present in the population means that its continuance into the next generation is a precarious business, and that it can easily be eliminated from the population by purely chance events.

Now consider the same allele at the same frequency but this time in a population of 10 000. In this population the allele will be represented 200 times, again mainly in the heterozygous state. Again there will be chance losses of the allele in the same way as before but in this case there are sufficient individuals carrying the allele for chance to act in both directions, that is, for as well as against the allele in question. Therefore in this case we may obtain a slight drift either side of the frequency of 0·01, but it is very unlikely that the allele would be eliminated from this population by drift alone.

8.3.4. *Selection*

(a) *General considerations.* In our consideration of factors affecting gene frequencies so far we have assumed that the members of a pair of alleles are of equal value to the organism. Thus if a gene exists as two alleles, for example *M* and *m*, then an organism's ability to survive and reproduce is identical, whether its genotype is *MM*, *Mm* or *mm*. The most important point is that for allelic frequencies to remain constant there must be no difference in the numbers of offspring, or their viability, produced by these three genotypes.

If a gene can be recognised as having two allelic forms *M* amd *m*, then clearly the gene must be responsible for a phenotypic character of one sort or another. Thus with respect to this gene there must be at least two, possibly three different phenotypes within a population. When the whole gene-complement of an organism is considered a massive reservoir of phenotypic variation is available. The likelihood that some members of such a population, who possess certain characteristics, will be at an advantage over the rest, forms the basis of Charles Darwin's theory of *evolution by natural selection.*

Darwin and A. R. Wallace, who came to similar conclusions at about the same time, presented a paper in 1858 outlining a mechanism by which evolutionary change could take place. They had no knowledge of Mendelian or population genetics, but so accurate were their observations and so precise their deductions that this later knowledge merely adds substance to their theories. The basis of the theory is that in a variable population some organisms are better fitted to the environment, and therefore are more likely to breed successfully and transmit their characters to the subsequent generation. Thus the characters they possess can be said to have been *selected for.* Similarly characters which do not confer any advantage, or may be positively harmful can be *selected against* because their possessors are not able to survive and breed as successfully. Over many generations changes will take place in a population which make it better adapted to survive in its environment. Such change, which may be over enormous lengths of time measured in millions of years, or over relatively short periods if the environment alters rapidly, forms the basis of evolution. The key step, that of species formation, will be considered at a later stage.

We are now in a position to think of selection acting not upon phenotypic characters but upon the frequency of the genes which control them. Returning to our simple example, if the allele *m* gives a phenotypic character which improves the chances of survival and reproduction of those organisms which possess it, then its frequency within the population will increase. There will, of course, be a corresponding decrease in the frequency of *M*. In brief, selection will act for or against an allele, increasing or decreasing its frequency. The probability is that very few alleles will be neutral in this respect, that is, neither selected for nor against, and therefore selection is a most important factor in the determination of gene frequencies.

The effects of selection will vary greatly according to the intensity of selection and the nature of the allele being selected. Let us take the extreme case of complete selection against a dominant allele, *M*. In this case all the homozygotes *MM*, and heterozygotes *Mm* will be prevented from breeding and thus the allele is eliminated in a single generation. Complete selection against a recessive allele is not quite so drastic. Consider a population in which the gene frequencies of *M* and *m* are both 0·5:

Gene frequency	$\frac{M\ (p)}{0{\cdot}5}$	$\frac{m\ (1-p)}{0{\cdot}5}$	
Genotype frequency	$\frac{MM\ (p^2)}{0{\cdot}25}$	$\frac{Mm[2p(1-p)]}{0{\cdot}5}$	$\frac{mm(1-p)^2}{0{\cdot}25}$
Phenotype frequency	$\frac{M-}{0{\cdot}75}$	$\frac{mm}{0{\cdot}25}$	

If selection against *m* is complete then none of the homozygous recessive organisms, *mm*, produce offspring. The effective breeding population thus consists of organisms *MM* and *Mm* which together have a frequency of 0·75. The frequencies of the two genotypes within this population thus becomes:

$$\text{frequency of } MM = \frac{0{\cdot}25}{0{\cdot}75} = 0{\cdot}33$$

$$\text{frequency of } Mm = \frac{0{\cdot}5}{0{\cdot}75} = 0{\cdot}66$$

If we assume these organisms mate at random and produce equal numbers of offspring, the alleles contributed to the subsequent generation can be calculated:

Alleles contributed to next generation

		M	*m*
Genotype	*MM*	2×0·33=0·66	
	Mm	0·66	0·66
approximate total frequencies		1·32	0·66

$$\text{frequency of } M = \frac{1{\cdot}32}{1{\cdot}32+0{\cdot}66} = 0{\cdot}66$$

$$\text{frequency of } m = \frac{0{\cdot}66}{1{\cdot}32+0{\cdot}66} = 0{\cdot}33$$

Thus in a single generation complete selection has reduced the frequency of the recessive allele, *m*, from 0·5 to 0·33. At the same time the proportion of organisms showing *m* phenotypically, that is those homozygous for it, will fall from 0·25 to 0·11 $[(1-p)^2 = 0{\cdot}33^2 = 0{\cdot}11]$. Thus the fall in the frequency of the allele is less rapid than the fall in the frequency of the homozygous recessive genotype. This is the explanation for Table 5 in which it is shown that the rarer a recessive allele is in a population the greater the proportion of heterozygotes to homozygotes. Thus in the case of rare recessive alleles continual selection over a long period of time is required if any significant change in the population is to be made. The existence of a fairly large number of lethal recessives in the genotypes of wild populations of many organisms indicates that even complete selection against a recessive allele need not necessarily eliminate it completely from a population.

We have so far considered selection only at its extreme value. In our examples an allele has been selected completely for or against. Such complete selection is relatively rare. It is more often the case that the possession of a certain allele confers a slight advantage or disadvantage in terms of viability, breeding success, number of offspring and so on. Selection of a lower intensity will still result in extensive changes in the frequency of genes within a population, but the number of generations, and hence the time involved, will be correspondingly greater.

(b) *Examples of natural selection. Industrial melanism in Biston betularia.* If natural selection is effective in altering the gene frequencies in wild populations, then one would expect its effects to be most noticeable in areas where there has been a pronounced change in the environment. Such a change will alter the selective pressures on the organisms and bring about rapid changes in gene frequency until a new balance of adaptation is achieved.

The onset and spread of the Industrial Revolution in Britain gave rise to a vast increase in the amount of atmospheric pollution. Huge quantities of soot deposited from the air caused blackening of buildings, trees, etc., altering significantly the environment of a number of organisms. Coincidental with this change in the environment it has been found that in over seventy species of moths black or darkened forms have become increasingly common. This phenomenon has been termed *industrial melanism*, and we shall take the single example of the peppered moth, *Biston betularia*.

Collections of *B. betularia* up to 1848 were almost entirely made up of individuals with a greyish-white colour speckled with black. In that year a black form, named *carbonaria*, was caught in Manchester. It differed from the normal form by a single gene, possessing the dominant allele *B*. This allele caused pigmentation by the deposition of melanin. The normal coloured moths are homozygous for the recessive allele, *b*. The melanic gene spread rapidly through the population until, by 1895, 99% of moths caught in the Manchester area were the melanic form.

It was suggested that this very rapid change in gene frequency was caused by a change in selection pressure brought on by the spread of industrialisation. *B. betularia* commonly rests, with wings outspread, on the bark of trees. In unpolluted regions the bark is often covered by a growth of lichens and in these circumstances the normal form is especially well camouflaged against predators, particularly birds. In similar circumstances the melanic form is very conspicuous, and would be heavily predated by the birds. Thus even if mutation continues to produce the melanic form, there would be heavy selection against it. One of the earliest effects of atmospheric pollution is to kill lichens, and when this is followed by a deposition of soot on the bark the camouflage situation is completely reversed. It is now the melanic form which is well camouflaged and the normal form which is conspicuous. Thus the direction of selection is altered and the allele *B* causing melanism is heavily selected for, and the normal allele *b*, selected against (Plate IX).

Elegant confirmation of this theory has been provided by the work of H. B. D. Kettlewell. He released marked moths, of both types, in two areas. One was a wood in a heavily polluted industrial region near Birmingham and the second an unpolluted wood in Dorset. Using a light moth trap he recaptured as many of the moths as possible, see Table 6.

TABLE 6 (after Kettlewell)

	Birmingham		*Dorset*	
	Light	*Dark*	*Light*	*Dark*
Number of moths released	64	154	496	473
Number of moths recaptured	16	82	62	30
Percentage recaptured	25·0	52·3	12·5	6·3

Thus in the soot-blackened wood in Birmingham the melanic form had a survival rate twice as high as the normal form. In the unpolluted Dorset wood the situation is almost exactly reversed.

At both sites birds were actually filmed taking moths from trees where equal numbers of the melanic and normal forms were released. The proportions of moths eaten at both sites are shown in Table 7.

TABLE 7 (after Kettlewell). Number of moths eaten when released in equal numbers

Birmingham		*Dorset*	
Light	*Dark*	*Light*	*Dark*
43	15	26	164

When offered the choice, therefore, birds predating on *B. betularia* will eat an excess of the form which is not camouflaged.

More recent research has shown the situation to be a little more complex. The proportion of melanic moths in a population seems to depend not only on the degree of atmospheric pollution, but also on its spread. Areas which are not themselves particularly industrialised often show very high proportions of melanic forms. Two climatic factors of great importance in this respect are the direction of the prevailing wind and the amount of rainfall. The former will determine whether atmospheric pollution is carried towards, or away from, a given area, and the latter the amount of soot, etc., deposited from the air.

Another peculiarity of industrial melanism in *B. betularia* is that, even in the most heavily polluted areas, populations consisting of 100% melanics are rare. It has been suggested that of the two genotypes which give melanic moths, *BB*, and *Bb*, the heterozygous form has some slight selective advantage. This of course, may be totally unrelated to the provision of camouflage, and is probably a physiological factor, such as rate of growth or viability of offspring. Selection, therefore, will operate to maintain an optimum frequency of the recessive allele *b* in the population, and this in turn will mean that a small proportion of light coloured moths, that is, the homozygous recessive, will continue to be found. This state of affairs, where two genetically controlled forms exist within a population, at frequencies determined by natural selection, is termed a *balanced polymorphism.*

Selection for heavy metal tolerance on mine waste tips

The piles of waste left after mineral extraction always contain considerable amounts of heavy metals. Occasionally there may be up to 1% of such metals present in the soil, but even using modern methods of extraction biologically active amounts still remain. The chief metals concerned are copper, lead and zinc all of which are extremely toxic in minute quantities to plant growth. In addition such tips are often very porous and contain very low levels of plant nutrients.

In spite of these extreme conditions several species of plants commonly grow on mine waste tips. Among those more frequently found are *Agrostis tenuis* and A. *stolonifera, Festuca ovina* and *F. rubra, Rumex acetosa, Plantago lanceolata, Anthoxanthum odoratum* and *Minuartia verna.* All these species grow readily on normal soils, so it is suggested that populations found on mine tips exhibit a *tolerance* to the metals present. Populations growing on normal soil do not exhibit such tolerance. This can be seen by growing the plants in question in culture solutions with and without the metal concerned. Plate X shows the root growth made by plants from tolerant and non-tolerant populations in a toxic solution. This, together with the fact that such tolerance persists after many generations in culture, indicates that tolerance to metals in these plants is a genetically controlled characteristic and, as such, susceptible to selection. Further information on the genetics of metal tolerance suggests that it is specific to a particular metal, for example tolerance to copper does not imply tolerance to zinc.

Tolerance to all these metals does, however, show a high degree of heritability, and is controlled by many genes which show a variable degree of dominance.

It is clear that on mine waste tips natural selection will act very strongly in favour of genes conferring tolerance to metals within the soil. Seeds from non-tolerant plants will germinate, but on the whole will make poor growth and the vast majority eventually die. Seeds from tolerant plants will germinate and grow to give normal adult plants. It is possible for natural selection to bring about the very rapid evolution of tolerant populations although probably in a majority of cases tolerant populations have been in existence for many thousands of years.

Professor A. D. Bradshaw (1970) has summarised much of this work on metal tolerance. He quotes the work of Snaydon who has shown that the plants under galvanised iron fences, which were erected in 1936, are tolerant to the zinc which washes off the wire.

A most interesting feature of these studies, particularly on *Agrostis tenuis*, is that between 0·3 and 0·4% of seeds from non-tolerant parents will germinate and grow on mine soil (Fig. 74). It can be shown that these plants exhibit a tolerance often

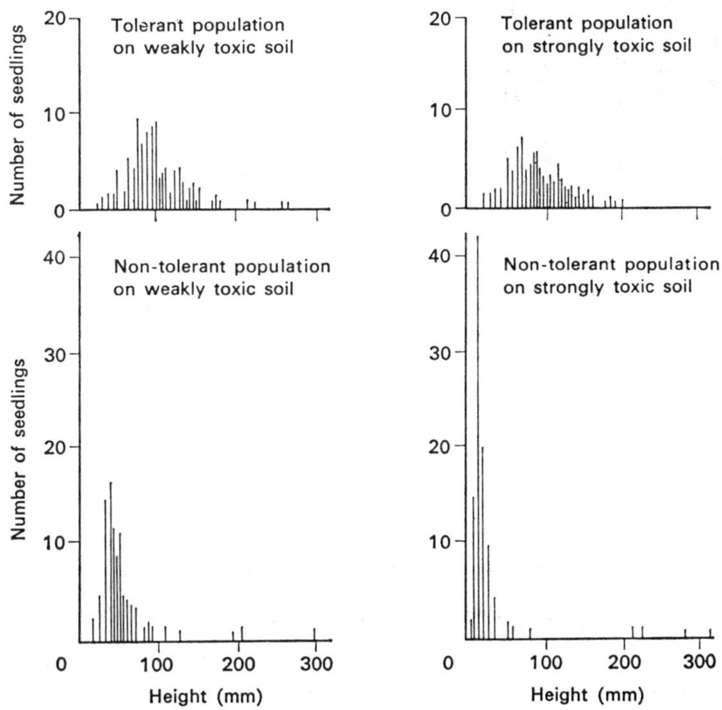

FIG. 74

Seedling height in tolerant (upper) and non-tolerant (lower) *Agrostis tenuis* populations grown in mixtures of 3:1 (weakly toxic, left) and 12:1 (strongly toxic, right) copper soil: John Innes compost equal to that of already tolerant material. This indicates that genes conferring tolerance exist in normal populations and that they can be exposed by the normal processes of recombination and reassortment. Once exposed such genes will be selected for in a single generation on mine soils.

Many of the species showing tolerance are outbreeding and wind-pollinated. It may be expected, therefore, that the resultant gene flow would tend to blur the distinction between tolerant and non-tolerant populations. Such is the power of selection in this case, however, that the effects of gene flow are frequently totally overcome. Tolerant and non-tolerant populations may exist within a few feet of each other, the transition from one to the other taking place just as rapidly as the change in the soil. Jain and Bradshaw (1966) have shown such a case at the Trelogan zinc mine (Fig. 75).

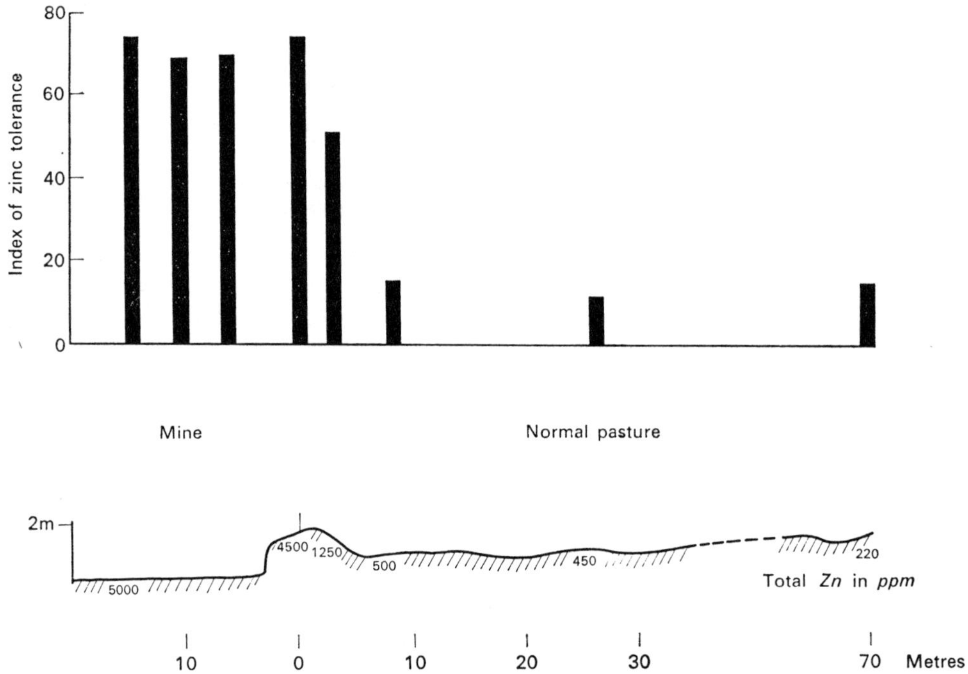

FIG. 75

Tolerance of *Anthoxanthum odoratum* at boundary of Trelogan zinc mine

Selection for shell colour and banding in Cepaea nemoralis

Cepaea nemoralis is a species of snail found commonly in central and western Europe. Its most striking feature is its variability. Apart from the lip which is almost always black or dark brown, rarely white, the shells vary in colour between yellow and brown and shades ranging from pale fawn, through pink and orange to red. In addition the shell may possess up to five dark brown or black bands running around the whorls of the shell. All the possible combinations of presence, absence and fusion of bands are found, although some types are much more common than others. The three most frequent banding patterns are shown in Fig. 76.

Unbanded 00000 | Mid-banded 00300 | Five-banded 12345

FIG. 76
Cepaea. Three common banding patterns

It is known that these characters are genetically controlled. For instance at one locus controlling colour, the pink allele is dominant to the yellow. At another locus the allele giving unbanded shells is dominant to the five-banded allele. The loci controlling shell colour and banding are known to be linked. Brown is dominant to the absence of brown, whether it be pink or yellow. The mid-banded character, see Fig. 76, is controlled by the action of a modifier gene on the five-banded allele, and is not linked to colour or banding.

Cain, Sheppard and a number of other workers have made huge numbers of collections of *C. nemoralis* from a wide variety of areas and habitats. Almost invariably, where the collection was of a reasonable size, they found that a population contained several types, or morphs, of snail with respect to shell colour and banding. The frequencies, however, of the various morphs differed greatly between populations. Cain and Sheppard have argued that in the Oxford district the polymorphic situation in this species is modified by natural selection, causing differences in gene frequencies between populations. The agents of selection are predators, chiefly the thrush, *Turdus ericetorum.* The various combinations of shell colour and banding in different populations offer camouflage against such predators on a variety of backgrounds. They sampled snails from Marley Bog, a small fen near Wytham Woods, Oxfordshire. The snail colony is predated by thrushes which carry the snails to a nearby bank where they break open the shells on protruding stones termed anvils. Comparisons were made between the number of live snails found and the number of broken shells over a period of sixteen days. They were classified into those shells which were 'effectively banded', that is, had bands visible to a predator and 'effectively unbanded', that is, did not have visible bands.

TABLE 8 (after Cain and Sheppard *Genetics* vol. 39, no. 1, 1954)

	Effectively unbanded	*Effectively banded*	*Totals*
Living snails	296	264	560
Predated snails	377	486	863

These results suggested that the effectively unbanded snails at Marley Bog were at a selective advantage compared with the effectively banded snails. These snails were presumably better camouflaged against that particular background than banded snails, and therefore were picked out less frequently by the thrushes.

Sheppard has also demonstrated selection for colour by thrushes. He has shown in collections from Wytham Woods that when in early spring the woodland floor is brown due to leaf litter and exposed earth, yellow snails are at a disadvantage compared with pinks and browns. The proportion of broken yellow shells at anvils being higher than the proportion of yellow snails in the whole population. However, later in the year when the ground cover becomes greener the situation is reversed and the yellow shells are at a selective advantage over pinks and browns.

If collections of *C. nemoralis* from lowland areas are classified according to the exact type of vegetation background upon which they are living, a strong correlation is seen between the frequencies of the morphs in the population and the background. For example the greener the background the higher the proportion of yellow snails, the more uniform the background the higher the proportion of unbanded snails (Fig. 77). The aggregation of samples from similar habitats on this scatter diagram indicates that morph frequencies within a sample are determined by visual selection by predators such as the thrush.

Since a particular colour banding combination confers camouflage against a certain background we would expect samples from similar habitats to show similar morph frequencies.

The situation in *C. nemoralis* is similar to that in *B. betularia* in that alleles are rarely completely absent from a population even when at a selective disadvantage, that is, the species exhibits a balanced polymorphism. As in *B. betularia* it is suggested that the polymorphism is maintained by some form of selection opposing the tendency of visual selection in favour of cryptic forms to break it down.

8.4. ISOLATING MECHANISMS AND SPECIES FORMATION

We have seen that several agencies are able to bring about changes in gene frequencies within a population. Of these agencies natural selection is almost certainly the most

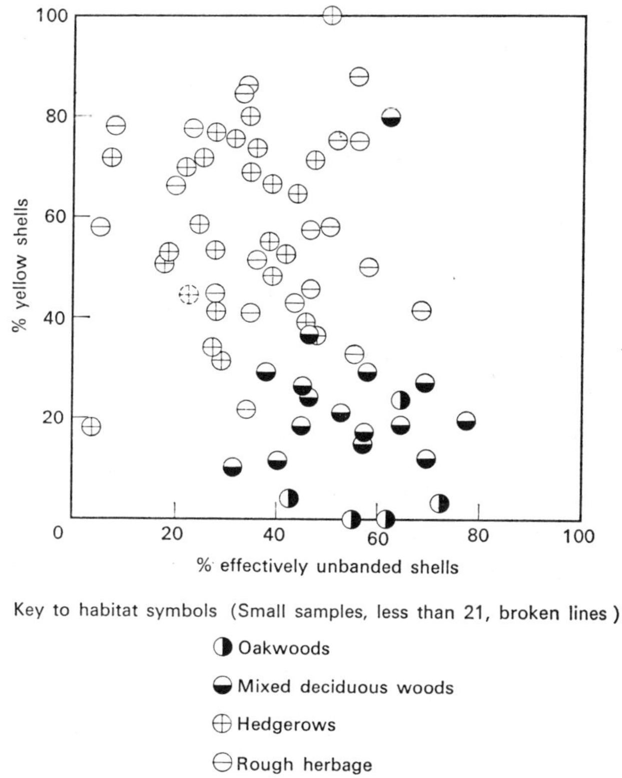

FIG. 77

Scatter diagram for percentage yellow shells, percentage effectively unbanded shells and habitat, of samples of *Cepaea nemoralis* from localities within 10 miles (16 km) of Oxford

important in wild populations. By responding to the selective pressures exerted by the environment a population becomes better adapted to that environment, that is, change takes place in the genetically controlled characteristics of the population.

The nature of the change will, of course, be dependent upon the precise selection exerted by the environment, and therefore may well differ between populations which are living in slightly different environments. Indeed the changes brought about by natural selection are likely to be different within a population, particularly if it is large, widely dispersed and divided into sub-groups. While such sub-groups within a population are able to freely interbreed, changes in gene frequencies taking place in different directions will be masked. In turn this means that there can be no permanent differentiation between the sub-groups as the gene flow between them will tend to counteract the individual changes in gene frequencies. The degree of adaptation of a sub-group to its environment which natural selection can cause is therefore restricted by the free flow of genes between it and other sub-groups upon which selection may be acting in a different direction.

Further, permanent adaptation can only occur when the sub-group becomes reproductively isolated from all others. Such isolation can occur by a number of mechanisms, but in all cases in effect what happens is that a barrier is erected between previously interbreeding groups preventing gene flow. Selection is now free to bring about changes in the sub-group which are permanent because they cannot be masked by gene flow from other groups. In these circumstances individual groups are free to change completely independently of each other giving a radiation away from the original population, each line being adapted to its particular environment,

The groups of organisms formed by such a process will eventually become distinct from one another and unable to interbreed to form viable fertile offspring. In other words each such group constitutes a *species*. Species formation brought about by reproductive isolation is the key process of evolution. It is an irreversible change. It allows changes in gene frequency to take place which are totally unrelated to changes in other species. Natural selection is able to act upon the gene frequencies within the species to bring about greater adaptation of that species to its environment.

While it is not an aim of this text to give a detailed account of evolutionary processes we feel it is essential to point out that changes in gene frequencies, together with reproductive isolation, form the basis of evolutionary change. Changes in gene frequencies are chiefly brought about by selection, but the effect of selection upon a group of organisms is limited until that group becomes reproductively isolated.

8.4.1. *Isolation by physical barriers*

Possibly the simplest form of isolation is that brought about by geographical barriers, such as seas, mountain ranges, deserts and so on. Such an isolating mechanism is passive, and non-genetic. Selection acting upon two groups isolated by a geographical barrier may produce two entirely different forms which are sometimes referred to as *allopatric species*. Speciation can only be said to have taken place, however, if the two groups cannot interbreed even if the geographical barrier is removed. In other words the original physical isolation must eventually lead to genetically controlled reproductive barriers for speciation to occur. Once a genetically controlled isolating mechanism is established, the need for geographical isolation disappears and the groups may coexist together. Closely related species of this type are termed *sympatric species*.

8.4.2. *Isolation by genetic barriers*

(a) *Habitat isolation*. If closely related species occupy totally different habitats this may well provide a reproductive barrier. In the Southern States of America two closely related species of frog, *Rana grylio* and *R. areolata*, are isolated by completely different habitat preferences. The former is aquatic and is found in deep ponds, lakes and so on, and breeds in deep water. The latter occupies disused mammal burrows during the day but is active at night around the edges of swamps. It breeds in isolated, shallow ponds. Thus the occupation of different habitats leads to reproductive isolation.

(b) *Seasonal isolation.* Occasionally closely related species become sexually mature at different times. Two species of lettuce, *Lactuca candensis* and *L. grammifolia*, both have haploid chromosomes numbers of 17 and are able, under unusual circumstances, to produce fertile hybrids. Under natural conditions, however, the former flowers in summer, the latter in early spring thus preventing hybrid formation.

(c) *Sexual isolation.* A variety of isolating mechanisms may be included under this heading. In a large number of animals mating is preceded by complex ritual courtship displays during which precise 'signals' are exchanged between participants. These may be in the form of movements, scents, or external markings the patterns of which are characteristic of a particular species. Thus a behavioural mechanism acts very strongly to isolate species.

Mechanical factors are also important in bringing about isolation in many plants and animals. In many flowering plants there is a complex relationship between the plant and an insect pollinating agent, which reduces the possibility of cross-pollination. In animals incompatibility between closely related species is often brought about by differences in the size and structure of the genitalia, a particularly important mechanism in Arthropods.

Gametic incompatibility is perhaps the most important factor in sexual isolation. In Chapter 7 the genetic basis of pollen incompatibility in a number of flowering plants was described, such a method proving a most efficient isolating mechanism. In a large number of animals the same effect is seen. The bullfrog *Rana catesbiana* and the closely related bronze frog *R. clamitans* fail to produce viable eggs, even when artificial crosses are made using sperm and ova preparations.

(d) *Hybrid inviability or sterility.* Where inter-specific matings do succeed the hybrids formed are very often weak and/or sterile. Hybrid seeds produced by two species of flax plant, *Linum austriacum* and *L. perensi*, fail to germinate. The mule, hybrid of the horse and the donkey, is sterile. Crosses between domestic cattle and yak or bison produce fertile cows, but sterile bulls. Hybrid inviability or sterility may be delayed for a generation, for example in crosses between certain species of cotton the first generation hybrids are vigorous, but the second generation are weak and largely sterile.

All the mechanisms described may be complete or partial and in most cases the exchange of genes is usually prevented by the co-operation of several mechanisms. Whichever mechanism is in operation it allows the species to diverge in isolation and become more closely adapted to its particular environment.

9
Applied genetics

THIS chapter is directed along two main lines of thought. The first part deals with aspects of medical genetics; the second with the genetics of crop plants and domesticated animals. Both sections are basically a collection of examples with no continuity between individual parts of each section, although where possible the examples are presented in increasing complexity.

9.1. HUMAN GENETICS

Although the science of genetics has mushroomed over the past century, its application in man has taken a much longer time to produce results. However the medical importance of genetics has, in the past decades, stimulated an acceleration in the gathering of information about hereditary conditions in man. Some of the instances detailed are innocuous variations such as the ability to roll one's tongue, the genetics of eye and hair colour, but naturally enough, much of the information which has been gained on human genetics is concerned with the inheritance of disorders.

9.2. CONDITIONS DUE TO SINGLE GENES

We have already mentioned a number of conditions present in man which are due to the mutation of single genes; such diseases as alkaptonuria and phenylketonuria (Chapter 2) are both caused by a recessive mutated gene. In both instances the mutation results in the loss of an enzyme involved in one particular metabolic pathway. The shape of the ear lobe, the ability to taste P.T.C., red-green colour blindness are all determined by single genes.

9.2.1. *Sickle cell anaemia*

This disease is caused by a recessive mutated gene which is involved in the construction of haemoglobin. This protein molecule has a primary structure consisting of chains of linked amino acids. At one point in this chain a molecule of the amino acid glutamine is replaced by the amino acid valine. This fault in the sequence is ordered by the mutated gene, Hb^S, forming abnormal haemoglobin S. This form of haemoglobin has different physical properties to the normal type; under low oxygen tensions it becomes increasingly insoluble and then physically distorts the red blood cells giving a typical sickle shape to the erythrocytes.

If the gene is present in the homozygous state $Hb^S\ Hb^S$ then all of the haemoglobin formed by the person will be the abnormal type. This usually results in the early death of the person from anaemia. In the heterozygous state $Hb^A\ Hb^S$ the Hb^S allele dictates the formation of haemoglobin S and the normal allele instructs the formation of the normal haemoglobin. This means that the effect of the haemoglobin S is diluted by the presence of the normal form. It is possible to determine whether an individual is heterozygous for this gene by chemically removing the oxygen from a sample of his blood. If the haemoglobin S is present a tendency to form sickle cells shows itself, with some of the red cells becoming sickled. However, the carrier himself shows no ill-effects, in fact to the contrary, he has a selective advantage. The proportion of heterozygotes in a population is almost nil in areas where there is no malaria, but in areas of the world where this disease is prevalent the frequency of the heterozygotes rises. Following this discovery it was found that the heterozygotes showed increased resistance to the disease.

The reason behind this resistance is still not clear but two possible explanations have been suggested. One is that the haemoglobin S is not readily metabolised by the parasite, the other is that the presence of the parasite in the erythrocyte lowers the oxygen tension causing the haemoglobin S to precipitate and the cell to sickle, and then be destroyed. Both lead to the demise of the parasite.

9.2.2. *Huntington's chorea*

This disease is caused by a dominant gene. The abnormal allele usually begins to show its presence when the person is between twenty-five and forty years old. The characteristic symptoms are repeated involuntary movements, violent twitchings and eventually mental disorder, all caused by the degeneration of nervous tissue. Unfortunately the disease usually develops after the person has had children of his or her own. These children then stand a one-in-two chance of inheriting the disease. Other diseases caused by dominant genes are gout, caused by an increase in the uric acid level of the blood, and achondroplasia which results in the dwarfism shown by some circus clowns.

9.2.3. *Sex-linked inheritance*

Both haemophilia and red-green colour-blindness (Chapter 5) are conditions caused by mutant genes carried on the *X* chromosome. In males, because of the absence of a second *X* chromosome and therefore any chance of a dominant allele, the recessive allele can express itself. Females rarely show the symptoms as the likelihood of both parents donating an *X* chromosome carrying the mutant allele is remote. The homozygous condition is therefore rare.

Muscular dystrophy is a severe disorder of muscles affecting approximately one boy born in every twenty thousand. The disease is caused by a recessive gene on the *X* chromosome. Until recently the detection of women, who were carriers of the allele, was not possible. However, it has been found that the amount of a muscle enzyme, creatinine kinase is higher in the blood of women who are carriers. By determining the amount of this enzyme it is possible to detect 70% of the carriers. With this type of information guidance can now be given on the chances of a woman giving birth to affected children.

9.3. VARIATION CAUSED BY MULTIPLE ALLELES— THE ABO BLOOD GROUP

This blood group was the first to be discovered. Landsteiner, at the beginning of this century, found that when the serum (blood plasma without the fibrinogen) from one man was mixed with the blood from other individuals one of two things occurred. Either the blood cells were unaffected or they clumped together. This clumping, called *agglutination*, is due to the presence of an antibody/antigen system. Antibodies are proteins, in fact part of the gamma-globulin fraction of the plasma, and these are made by the body to react with specific proteins which are foreign to the body. These foreign proteins, called antigens may be free proteins introduced into the body, such as the toxins from bacteria, or they may be part of a complex unit such as the cell membrane about protozoan parasites or the foreign erythrocytes. Usually the body has to be primed by an initial introduction to the antigen before it is able to synthesise an antibody which will react with that particular antigen. Two antigenic substances are involved in the ABO blood group system. These are antigens A and B and they form part of the erythrocyte membrane. Four types of phenotype are found, those possessing both A and B antigens (in group AB), those possessing either A or B and those possessing neither A nor B (group O). The ABO blood grouping is unusual in that antibodies specific to the *missing* antigens are already present in the plasma without prior exposure to the antigen. Therefore a person of blood group A will have in his plasma, antibody anti-B; a person of blood group O will have both anti-A and anti-B antibodies present and so on (see Fig. 78). If incompatible blood groups are mixed the antibodies will attach themselves to their specific antigens, linking adjacent erythrocytes together into groups and producing the characteristic agglutination.

The inheritance of the ABO blood group is controlled by the multiple allelic condition of the gene L (after Landsteiner). Allele L^A will initiate the formation of antigen A. Allele L^B will initiate the formation of antigen B. A recessive allele l also exists. Six allelic pairs are possible:

$\left.\begin{array}{l} L^A L^A \\ L^A l \end{array}\right\}$ will produce antigen A

$\left.\begin{array}{l} L^B L^B \\ L^B l \end{array}\right\}$ will produce antigen B

$L^A L^B$ will produce both antigen A and antigen B

$l\,l$ will form neither antigen

Knowledge as to which blood group a person belongs is of importance in blood transfusions. During a transfusion it is necessary to ensure that the donated red cells are not agglutinated by the patient's antibodies. The relatively small quantity of antibodies

Blood group	Antigens on erythrocytes	Antibodies in the plasma
A	A	anti B
B	B	anti A
AB	both A and B	neither
O	neither	both anti A and anti B

FIG. 78
Summary of the antigens and antibodies involved in the ABO blood group

in the donor's serum will be rapidly diluted by the large volume of the patient's blood and the effects of the introduced antibodies will be reduced. People of blood group O make good donors as their red cells carry no antigens and therefore cannot be agglutinated; they do, however, make poor recipients as their serum contains both antibodies and will thus agglutinate donated blood of A, B and AB groups. The table in Fig. 79 extends and summarises this reaction between the donated red cells and the recipient's antibodies.

9.4. CHROMOSOMAL MUTATIONS IN MAN

9.4.1. *Down's syndrome*

This disease results in characteristic facial features which superficially resemble the Mongolian races, and is thus also called mongolism. Children affected by the disease have a round, slightly flattened face and head, with small slanting eyes. The bridge of

Donor	Recipient				
	Blood group				
Antigens on the erythrocytes	A	B	AB	O	
	Antibodies in the plasma				
	anti B	anti A	neither	both anti A and anti B	
A	–	√	–	√	
B	√	–	–	√	Universal recipient (AB column)
AB	√	√	–	√	
O	–	–	–	–	
	Universal donor (O row)				

√ Indicates agglutination

FIG. 79

Summary of possible blood transfusions

the nose is often absent or poorly developed and the ears are small. The fingers are short, especially the little fingers and there are abnormal palm prints. Varying degrees of mental retardation also occur, and many suffer from heart diseases. About 0·2% of children born in Europe suffer from the disease. The frequency of the syndrome increases with the age of the mothers; this means that a child born to an older woman is more likely to suffer from the disease than one born to a younger woman.

The syndrome is caused by the presence of an extra chromosome in the cells. It appears that the pair of chromosomes 21 fail to separate during gametogenesis. A gamete is therefore formed with a pair of chromosomes 21 instead of only one member of the pair. Fertilisation of this gamete with a normal gamete produces a zygote, and subsequently a child, with three chromosomes 21—the trisomic condition that is, the child will have a chromosome number of forty-seven, and not forty-six (see Plates XI and XII). This form of non-disjunction apparently increases in frequency in older women, thus increasing the risk of the formation of mutated ova. This in turn results in a greater proportion of mongol children born to older women.

Down's syndrome may also be caused by translocation (Fig. 80). Occasionally exchange of segments occurs between the chromosomes 21 and 15. This results in a

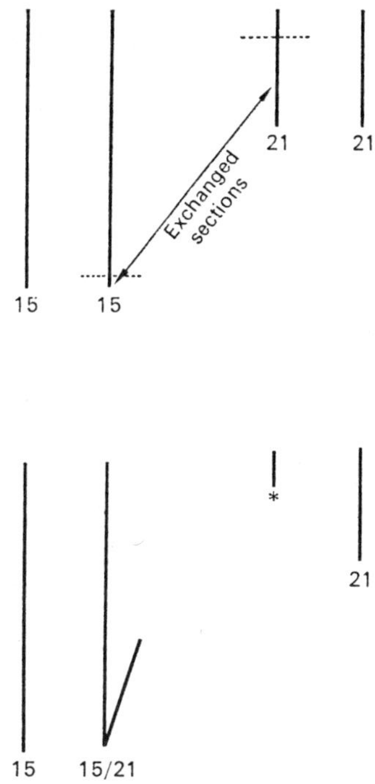

FIG. 80
Translocation between chromosomes 15 and 21

very long chromosome 15 which is carrying a large portion of chromosome 21. If this mutation occurs in the somatic cell division of an individual he will not be affected as there has been no loss or gain in the total genetic material of his cells. However, if it occurs during gametogenesis, he will form two types of gametes, some will carry one chromosome 21 and a normal chromosome 15, others will contain one chromosome 21 and an abnormal chromosome 15/21. This second type of gamete is in effect carrying two doses of chromosome 21. Fertilisation of this type of gamete by a normal gamete will form a child with a normal chromosome number, but with a triple dose of chromosome 21 and this child will develop the syndrome. Cytological examination would show a chromosome count of forty-six, but one of the chromosome 15 would be very long. This disease is therefore hereditary, the parent passing on the translocated material to the children (Fig. 81), and is responsible for family histories with recurrent cases of Mongolism.

9.4.2. *Klinefelter's syndrome and Turner's syndrome*

Klinefelter's syndrome is caused by an abnormality of the sex chromosomes, resulting in the trisomic condition *XXY*. Individuals of this genotype look and behave like men

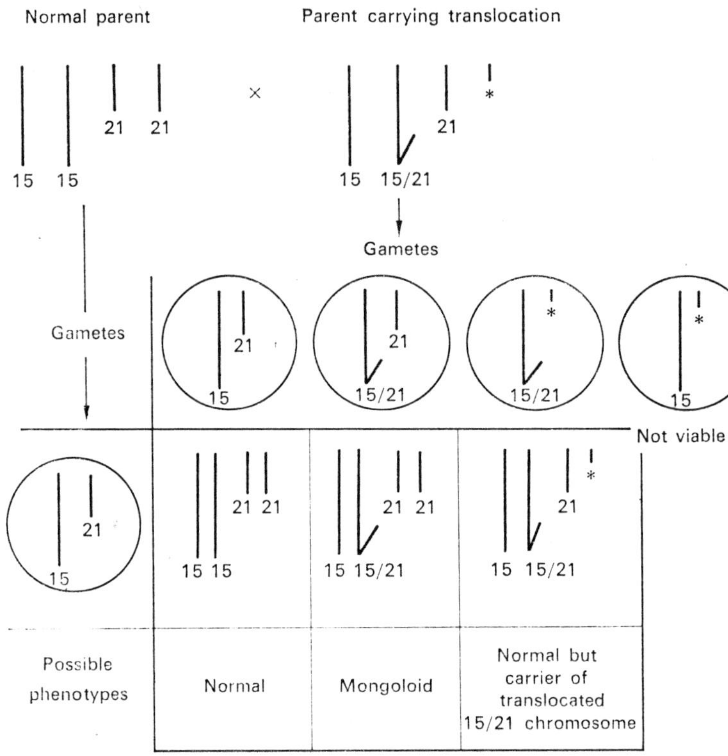

FIG. 81
The inheritance of translocated chromosome 15/21

but have a number of female characters such as high voice, absence of facial hair and some development of breast tissue. They are sterile, having small underdeveloped testes. The additional X chromosome is due to non-disjunction in one of the parents. If this form of aberration occurred during gametogenesis in the male two abnormal types of sperm would be formed, one would carry both sex chromosomes, XY, and the other would carry neither. In the woman non-disjunction would again give rise to two abnormal gametes, one XX the other again with neither sex chromosome. If an XY sperm fertilises a normal X ovum, the trisomic condition XXY would result, the same situation could arise if a normal Y sperm fertilised an abnormal XX ovum. Figures 82(a) and (b) show the possible combinations involving non-disjunction in the male or female.

Three other abnormal genotypes may also occur from such crosses. XXX individuals are female and although trisomic for the X chromosome are in fact normal. YO individuals are probably inviable. Evidence for this comes from studies of *Drosophila* where in the absence of the X chromosome the YO zygote dies very early in the egg

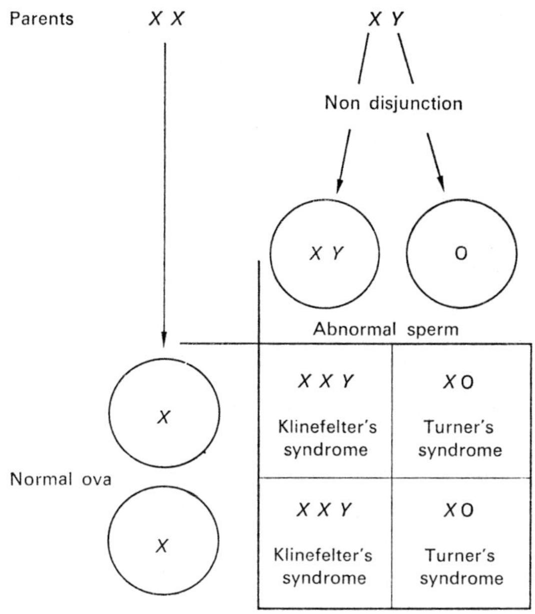

FIG. 82(a)
Non-disjunction of sex chromosomes during spermatogenesis

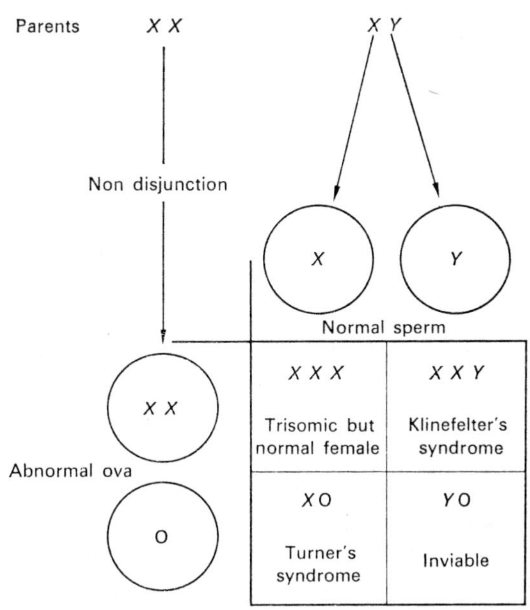

FIG. 82(b)
Non-disjunction of the sex chromosomes during oogenesis

stage, probably due to the absence of vital genes which are only represented on the *X* chromosome. *XO* individuals are monosomic for the *X* chromosome. Persons with this genotype are female but they suffer from a condition known as Turner's syndrome. The individual fails to mature sexually, is often short and often has a webbing of the neck due to the presence of extra skin between the neck and shoulders.

9.5. EUGENICS AND GENETIC COUNSELLING

Eugenics is a term coined by Francis Galton, it basically means the study of those social factors which improve or have deleterious effects upon the gene pool of the population. The science would apply artificial selection to the population by encouraging individuals with desirable characters to produce offspring, at the same time discouraging individuals with unfavourable characteristics from breeding. This sounds very simple and clinical but poses two main problems. We have already seen that phenotypic characters are rarely the result of the genotype alone, but that environmental factors act in combination with the genotype. Having determined the effects of both factors it is then necessary to decide which characters are desirable traits. Desirable traits are in fact difficult to determine, undesirable characteristics are far more obvious. Usually these undesirable traits are the result of single mutations which make them easier to investigate than the desirable characters which are often controlled by a whole range of genes showing complex patterns of inheritance. Many of the very harmful genes cause self-destruction because they are lethal or cause sterility in the homozygous state. However, methods are required by means of which the presence of these alleles may be detected in the heterozygous state. Given these methods it would then become possible to deter heterozygotes from marrying or having children, thus reducing the likelihood of the homozygous condition arising. As we have seen it is possible in the case of muscular dystrophy (Duchenne type) to detect by blood tests about 70% of the women carriers. A similar possibility exists in the case of the gene causing phenylketonuria. In the homozygous state the afflicted person is unable to metabolise phenylalanine which is normally converted to tyrosine. Instead some of the phenylalanine forms phenylpyruvic acid, large amounts of which are found in the blood, cerebrospinal fluid and in the urine. One of the results of the high level of concentration of this substance, or one of its derivatives, is mental retardation. The heterozygotes also show slight biochemical differences from the normal: their plasma contains slightly more phenylalanine than the normal person, but there is some overlap. However if a carrier is fed a standard dose of the amino acid, it is not removed from the plasma as rapidly as in the normal person. In this way the presence of the gene can be detected in the heterozygous state. The estimated incidence of this disease is about three cases in 100 000. A much more alarming frequency is found in fibrocystic disease of the pancreas. This disease is caused by a recessive mutated gene and is the most common of all diseases caused by mutant genes, being responsible for one death in every two thousand babies born. The disease causes the secretion of mucus which is abnormally viscous; this blocks the fine air tracheoles in the lungs and results in infection. The viscous fluid also interferes with the digestive processes. Another characteristic is the presence of large amounts of salt in the sweat. The high incidence of the disease, and thus the homozygous recessive condition (1 in 2000) means that about four to five per cent of the population must be carriers. This high frequency of the allele in the population suggests that the heterozygote must have some selective advantage over the normal condition but just what this is is not known. Another possible explanation for the high frequency is an abnormally high mutation rate, but this seems less likely. Because of the large number of heterozygotes, about one in twenty people, the desirability of detecting the presence of the allele and advising carriers of it on the likelihood of their having affected children, is obvious. It seems possible that the carriers can be detected by testing their sweat which shows a slightly higher concentration of chloride ions than the normal person's, although the concentration is much below that of the homozygous recessive condition.

These are a few cases where the presence of a recessive deleterious gene can be detected in the heterozygous state. With this situation the carriers can be warned of their condition and of the risks involved if two people with a similar condition have children. By this form of dissuasion the proportion of children born with the homozygous mutant condition should fall. Ideally from the eugenic point of view, if the carriers had no children, or even if their family size was reduced below the average then the frequency in the population of the mutant gene would decrease. Complete elimination would be impossible due to constant formation of new alleles by mutation.

9.6. ANIMAL BREEDING

Plant and animal breeding involves the practical application of the concepts of human eugenics to crops and domesticated animals. In these organisms artificial selection can be applied to increase the frequency of genes considered desirable, and reduce the frequency of undesirable genes. Unfortunately this artificial selection of the ideal characters meets up with an early stumbling block, that of variation caused by the environment, and not by the genetic material.

The earliest experiments on this phenomenon were carried out by a Dutch botanist, Johannsen (1911). He used the runner bean (*Phaseolus vulgaris*) and started these, now classical, experiments with a mixed bag of seeds. The runner bean is self-fertilising and therefore over the generations the homozygous state will arise in each bean, with different beans being homozygous for different combinations of alleles. From his beans he selected the largest, which were presumably homozygous for a large proportion of genes promoting increased seed size, and the smallest. He found that plants grown from the large seeds produced on average large seeds, and the small seeds grew into plants which produced on average small seeds. Each line however produced a range of seeds, that is the 'large' plant produced a whole range of 'large' seeds. When he selected the largest and the smallest from this range, and grew them he found there was no difference in average size between the two sets of progeny (Fig. 83).

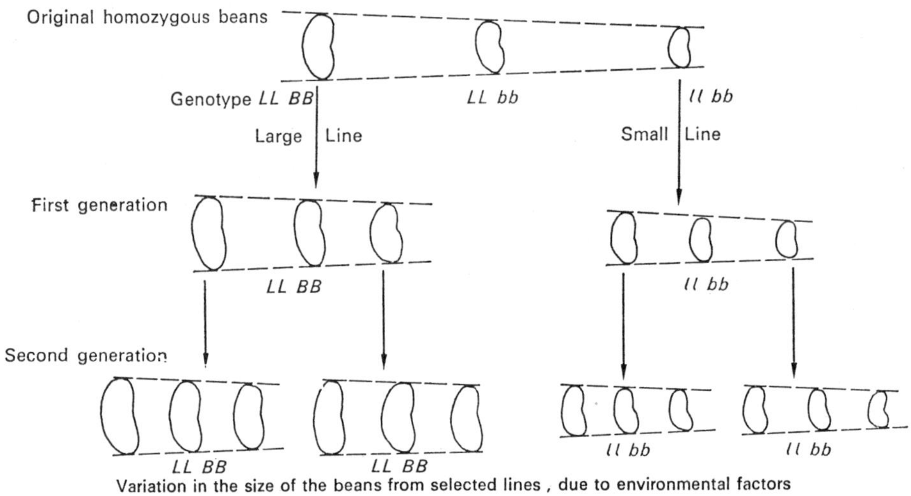

FIG. 83

Diagrammatic representation of Johannsen's experiments with beans

The explanation for the above differences is that the initial selection of large and small seeds was basically selecting two different lines of seed which differed genetically. However a homozygous plant with a 'large' genotype will give rise to offspring with the same homozygous condition, that is they will all be basically large. The observed variation in their size is entirely due to variations in the environment of the developing beans: such differences as their positions inside the pod in relation to the other beans and the position of the pod on the plant.

From the viewpoint of the animal and plant breeder the interference by the environment presents another problem, the exact effect of which is difficult to calculate. One important aid in this problem is the occasional appearance of identical twin animals. Twin animals can occur by two means: if two ova are fertilised at conception two embryos will be formed resulting in non-identical twins. Identical twins are formed when a single ovum is fertilised, and very early in the development of the embryo a split occurs forming two cell masses which both give rise to normal embryos. As both were originally derived from the same zygote they will have identical genotypes.

With identical twins, in man or other animals, both will have identical gene-determined characters, that is they will be the same sex, have the same eye and hair colour a 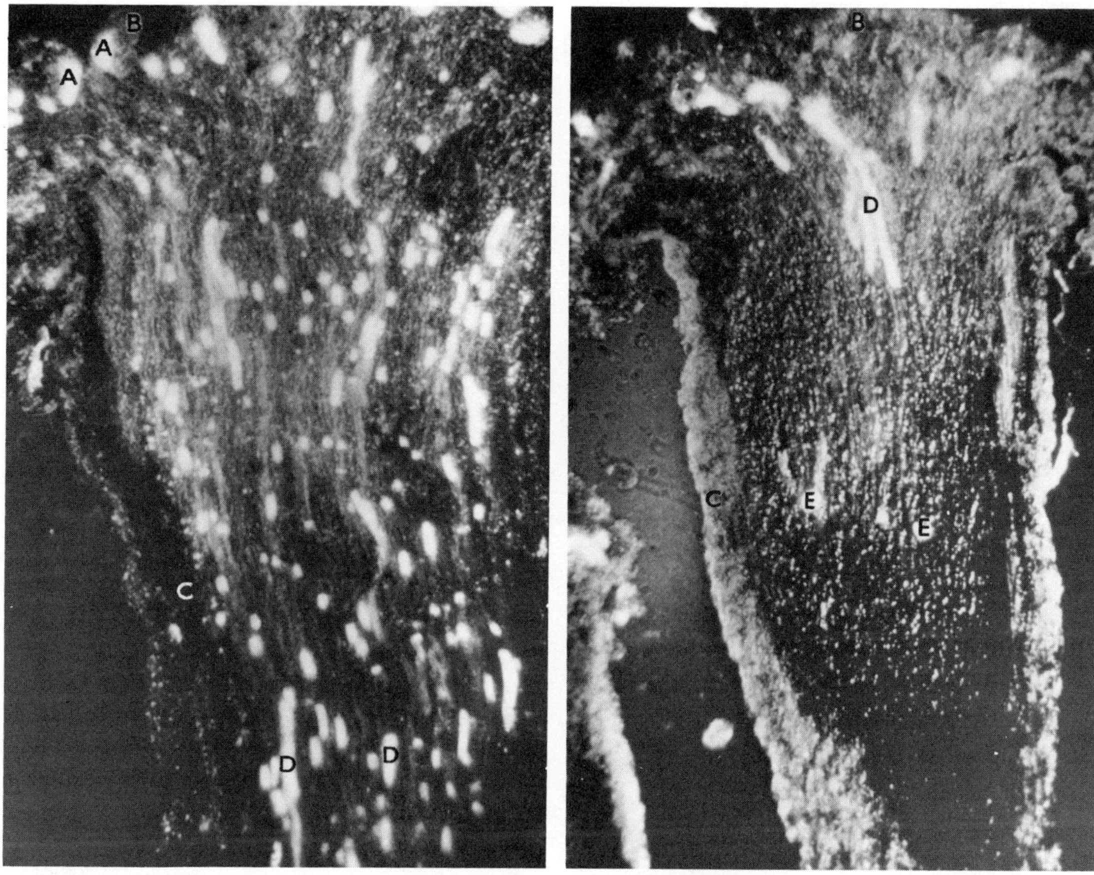 b

PLATE VIII a & b

Longitudinal sections through the stigma and upper portion of the style of flowers three days after pollination. Stained in aniline blue and viewed through transmitted u.v., reflected red light illumination.

a This flower was open pollinated. Pollen germination was good and the stylar canal contains numerous pollen tubes. These are indicated by fluorescent deposits of callose, the amount of callose in each tube varying inversely with the rate of growth of the tube.

b A self-pollinated Worcester Pearmain flower. This variety is self-incompatible. The pollen tubes contain heavy deposits of callose indicating a slow rate of growth. The tubes soon stop growing and form a bulbous tip of callose.

A Pollen grains
B Stigmatic surface
C Style
D Fluorescent callose in pollen tubes
E Bulbous tips of incompatible tubes

PLATE IX
Biston betularia, normal and melanic forms.

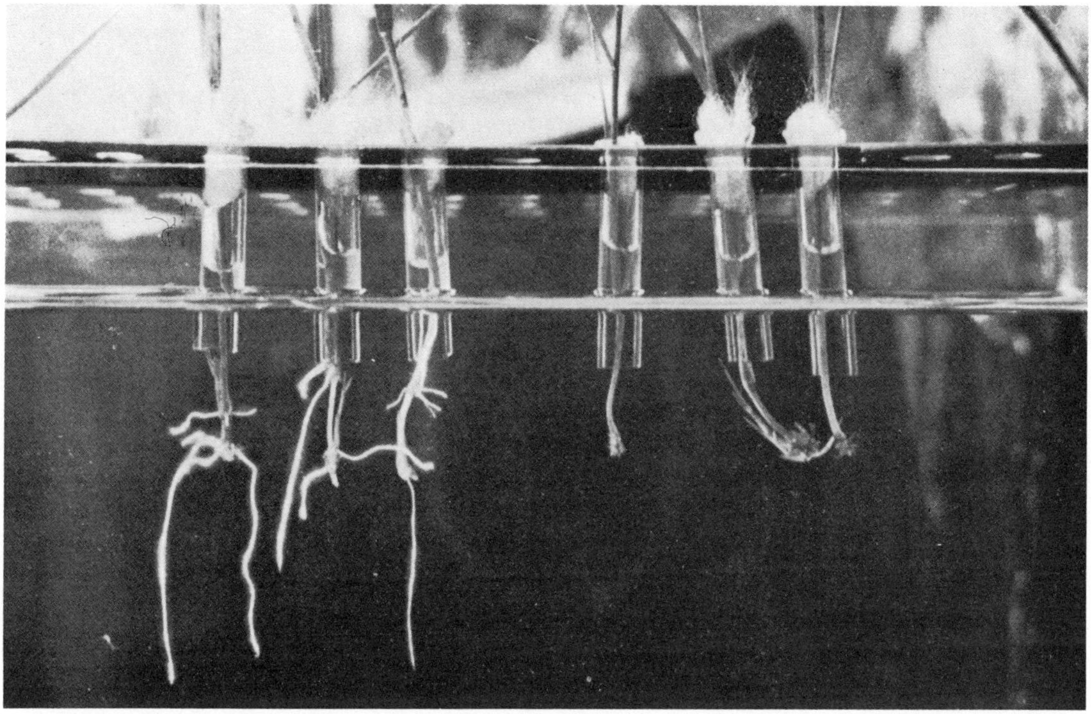

PLATE X

Root growth of tolerant (*left*) and normal, non-tolerant (*right*) populations grown in toxic solution.

a

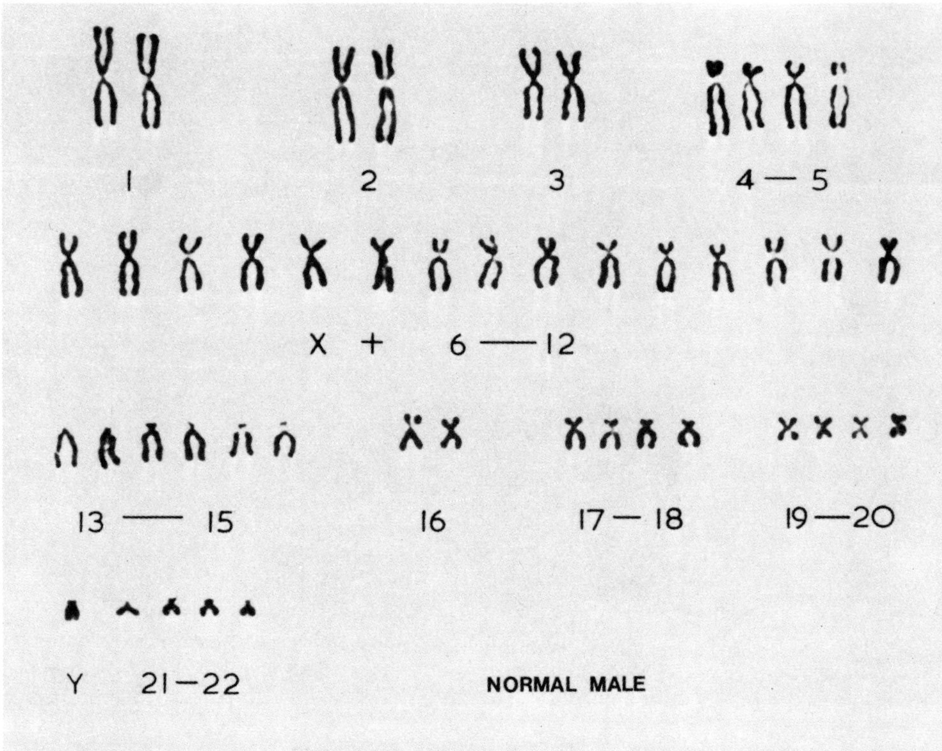

b

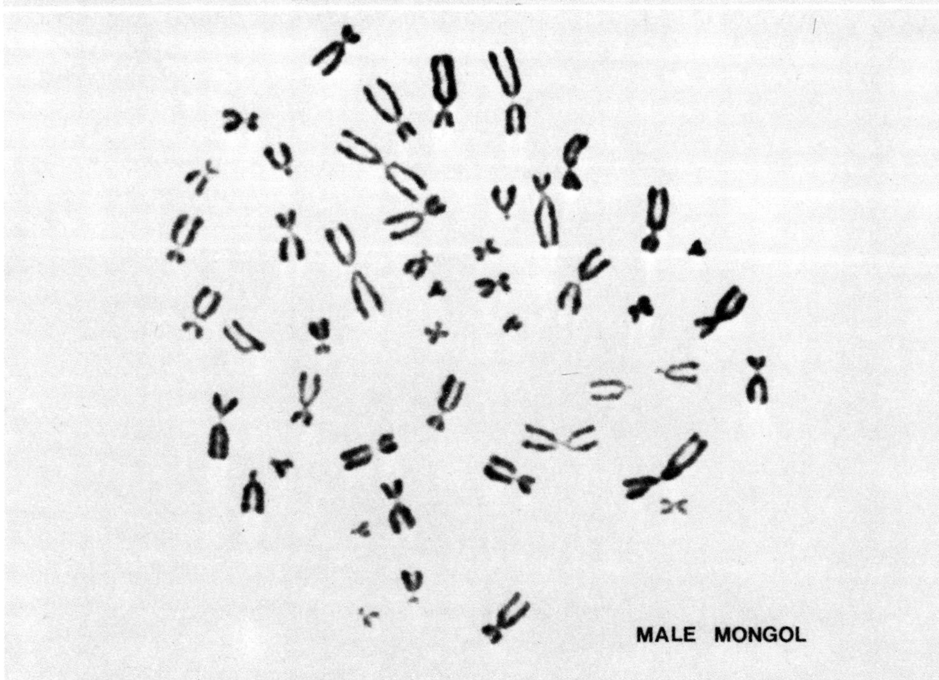

PLATE XI

a The normal human male karyotype.

b The chromosomes of a male mongol.

they will even have similar blood groups and fingerprints. In so far as characteristics modified by the environment are concerned they will differ, if the difference is great this will give an indication of the importance of the environmental factors in determining that character. Thus it has been found that identical twins who are separated at birth and reared by separate foster-parents under different environmental conditions differ strikingly in their weights although their overall heights and appearances are similar. This suggests that a characteristic such as weight depends to a great extent on the environment for its expression. Facts such as this are obviously important in animal breeding and twin experiments with cattle have borne this out.

9.6.1. *Twin cattle experiments*

Bonnier and co-workers at the Institute of Animal Breeding at Wiad, near Stockholm, carried out experiments on identical twin cows in an attempt to find out the relative effects of the genotype and environment on different features of the animals. A pair of identical cows (Sara + and Sara −) were separated: one was fed a minimal diet (Sara −) the other was fed with an abundance of food (Sara+). Their increase in weight was noted. The experiment was repeated using a second pair of identical twins from a different line of cows to the first pair. Spetsa + was fed with ample food, Spetsa − with the bare minimum. At the end of the experiment just over two years later Sara + weighed nearly 500 kg whereas her identical twin only weighed 360 kg. The Spetsa pair also showed this type of difference with the well-fed twin weighing 390 kg compared to her sister's 300 kg on the minimal diet. These differences, in genetically identical pairs, show that the amount and quality of the food does play an important part in the weight of the animal. But another revealing fact is also present. Both Sara + and Spetsa + were fed the same amount and quality of food, and yet they differed enormously in their final weights. This variation was most probably due to differences in the genotypes of the two lines of cattle. The Spetsa animals were genotypically inferior to the Sara twins and given similar foodstuffs the Spetsa + animal was only just able to increase in weight above the Sara − animal fed on a minimal diet. This suggests that the genotype sets the maximum limit for a particular character but the environment determines the exact placing below that limit. Similar results were also obtained with milk yield, but the fat content of the milk was less sensitive to environmental changes, being controlled more strictly by the genotype.

9.6.2. *Cattle breeding and artificial insemination*

Artificial Insemination (A.I.) has made possible rapid progress in animal breeding. In cattle breeding, because of the effects of the environment on the beef yield and milk yield of a particular animal, it is difficult to select for certain animals showing favourable characteristics which are genetically controlled. One reasonable method for selecting a specific animal is by using the progeny testing method. This involves examining the progeny from a given animal and if the overall quality of the factors under consideration is of the standard required, selecting that parent animal for breeding. Because of the small number of offspring a cow can produce it is not practical to judge the quality of the cow by progeny testing. The bull however can sire a much greater number of progeny and it is the quality of the bull which is usually assessed by this method. One character which has to be assessed by examining the progeny of the bull is its ability to sire good milk producers. Prior to A.I. the large number of daughters needed for progeny testing had to be produced by natural insemination which meant that by the time sufficient animals had been sired, had grown and then their milk yield had been found, the bull was generally too old for further successful breeding. However, the use of A.I. means that at organised centres semen, from good pedigree bulls, can be obtained, diluted and stored. Because the semen can be diluted many more cows can be inseminated, which in turn means that a new bull can sire a large number of progeny very quickly. The overall result is that progeny testing can be applied and records obtained before the bull is very old. Sperm from proven sires can then be stored in deep freeze for many years. With the introduction of the service farmers no longer have to keep their own mediocre quality bull, for bulls of far better genotype can be kept at the centres; the only expense to the farmer is the cost of each insemination. It also means that farmers can select the particular qualities they want in their stock from the wide range of semen stored at the centres.

There are, however, inherent disadvantages with this system. Great care has to be taken to keep in the centres as wide a selection of bulls as possible, in this way the variety of the gene pool will be maintained. To counteract inbreeding, the bulls usually stay at a given centre for a limited time only, usually about two and a half years; they are then moved to a different centre. Artificial insemination has also facilitated line breeding in animals. This involves the fertilisation of successive generations of cows with semen from the same proven sire. Cows from each generation are selected for the desired qualities and then basically inbred, in an attempt to accentuate that particular quality.

9.6.3. *The control of pests and diseases*

A second line of concern for the geneticist is the increasing number of animal pests and disease-causing organisms which are becoming immune to chemicals used to control them. This is a form of adaptation by an organism which has been exposed to an artificial environmental factor in the form of antibiotic or pesticide. Organisms which exhibit this form of adaptation are those which have a very high reproductive rate. This in turn means that with the large numbers of offspring produced, the infrequent favourable mutation is more likely to show itself and gene combinations conferring resistance are more likely to arise by recombination and reassortment.

Bacteria and viruses reproduce extremely rapidly under favourable conditions. In hospitals where the patients are being treated with antibiotics the selective pressures on the bacteria are high, and this results in the appearance of resistant strains which were already present as mutants in the bacterial population. There exist today strains of bacteria resistant to all antibiotics in use, likewise strains of insect pests are appearing which are resistant to various insecticides. Recently rats have shown similar capacities even though their reproductive potential is far below that of the insects: strains of rats are appearing which are resistant to the 'Warfarin' type of rodenticide.

Resistance in mosquitoes

These insects are important vectors of parasites which cause diseases in man. There are three main genera: (i) *Anopheles*, members of this genus are responsible for transmitting the protozoans responsible for malaria (*Plasmodium* spp), (ii) the *Culex* genus which contains mosquitoes responsible for transmitting the filarian nematode responsible for causing elephantiasis, and (iii) *Aedes* spp, members of which are vectors of the virus causing yellow fever. Because of their importance as vectors of these diseases insecticides, particularly D.D.T., have been used persistently in an attempt to eliminate these insects. Strains resistant to this and other insecticides are now very common. Two possible reasons could explain the appearance of such resistant strains. If the D.D.T. was mutagenic it might have induced mutations in the genetic material resulting in resistance to the insecticide. There is no evidence to support the view that D.D.T. is a mutagen and therefore it is most probably simply selecting insects which carry favourable mutated genes already in the population. Resistance to D.D.T. and related compounds is in fact due to a single recessive or semi-recessive gene. Resistance to insecticides in general is not the responsibility of a single gene, instead resistance to different insecticides is controlled by several different loci. For example the dominant or semi-dominant gene responsible for resistance to the Dieldrin type of insecticide is at a different locus to the gene responsible for D.D.T. resistance. Thus the introduction of a powerful new selective agent into the environment of these insects has led to the rapid spread of alleles conferring resistance. Clearly this type of control of pests and diseases can only be effective in the long term if the production of a new chemical agents can keep pace with the appearance of mutant alleles giving resistance.

9.7. PLANT BREEDING

Early experiments in plant breeding centred about the formation of more productive crop plants by mass selection or line breeding. Mass selection involves the growing of large numbers of the plants under consideration, followed by selection of individuals with favourable characters. Seeds from these selected plants are grown and selection of the next generation repeated. Darwin in his *Origin of Species* described how pear plants had been treated in this way to produce a much better variety than the original stock. Line selection involves raising plants from seed obtained by self-fertilisation of the parent plant. The plants so formed would be selected and those showing the desired traits again self-fertilised. The selected characteristic however favourable would also be accompanied by increasing degrees of homozygosity. This would mean that deleterious genes, which up to this point had remained ineffective due to the heterozygote condition, could now express themselves. The overall effect is that although line breeding may encourage the selected characters the accompanying deleterious effects would also show themselves.

A further development in plant breeding came in 1906 when Nilsson-Ehle published a paper in which he pointed out the faults of the line system method of breeding He went on to suggest that instead of selfing plants with favourable characteristics much better results could be obtained by artificially crossing two different lines of such inbred plants. The plants formed by this outbreeding derive the favourable characters from both parents and also show the return of hybrid vigour. By the same argument plants will also be formed which have all of the unfavourable characters from both parents, and these have to be discarded.

9.7.1. *Hybrid maize*

The above modification of line breeding has been used with great success in America in the production of hybrid maize. Parent plants were obtained by inbreeding several lines of maize. The resultant plants were homozygous for many genes and phenotypically showed great loss in hybrid vigour, that is they were smaller and weaker than the original plants, they were more difficult to grow and produced smaller and fewer 'seeds'. The separated lines of inbred strains are very different from one another but are very uniform within the line. This means that inbreeding has produced a state where each line is homozygous for most of its genes, the different lines being homozygous for different combinations of genes. A large number of such genes will be involved in the overall concept of the plant's vigour.

By crossing two plants from different lines (A and B) an *F1* generation of plants was formed with renewed hybrid vigour and very much increased yield. The overall effect of this heterosis was to produce plants which were often much more productive than the original cross-pollinated varieties. When hybrids from an A/B cross were then intercrossed with hybrids from a C/D cross (C and D representing two further lines) a further increase in the hybrid vigour and yield of the *F2* plants was obtained. The increased vigour of the *F1* and *F2* plants is due to the fact that any homozygous genes present which are 'anti-vigour' will be compensated for by the other lines 'pro-vigour' genes. This type of reshuffling in the genetic material, and selection of the favourable combinations, relies on the presence of genes which themselves are responsible for the favourable characters. By artificially inducing mutations in plants the plant breeder can produce even more variations.

9.7.2. *X-ray induced mutations*

Irradiation of plants using X-rays will induce genetic mutation; unfortunately the vast majority of these are unfavourable or harmful. A very small percentage of the mutants, however, might result in favourable changes in the plant, which can be selected for. This selection of X-ray induced mutations was used in barley and it has been possible to obtain such clear-cut benefits as earlier ripening, stiffer straw and a higher yield than in the original parent variety.

9.7.3. *Polyploidy in crop plants*

In 1916 Winkler and subsequently Jorgensen and Sansome showed that by decapitating a young diploid tomato plant, side shoots are formed which in many cases are tetraploid (4n = 48). These branches are easily recognised as they are stouter, the leaves are broader and a darker green and there is an overall increase in the dry weight of the fruit.

Artificial inducement of polyploidy can also be brought about by using colchicine or its derivatives (Chapter 6), although as with induced mutation by X-rays, only a very few of the polyploids formed are of any use and these usually have to be improved by further crossing. Such an example is the sugar beet (2n = 18) where induced tetraploids have been formed. These 4n plants, however, are inferior to the diploid plants, but the triploid form, obtained by crossing the diploid with a tetraploid plant, has a greater sugar yield than either parent. The triploid seed is obtained by planting the diploids amongst the tetraploids, producing a mixed seed, most of which is 3n with some 2n and 4n.

The cereal plants have also been subjected to this form of inducement, for example tetraploid rye plants have been produced which, although not as vigorous as the diploid strain, have a kernel which is 50% larger than the diploid parent's kernel. The 2n rye has also been crossed with wheat to form a hybrid plant, rye-wheat which has a very low fertility. This hybrid, on exposure to colchicine, doubled its chromosome number forming an allopolyploid variety which is fertile. These allopolyploids have been improved to the point where they yield 90% more than a good wheat variety. Other allopolyploids of this type have been produced with different favourable characteristics such as stiff straw, resistance to cold and disease, and research is now attempting to produce a combination of these lines with all of their advantages.

10

Practical methods

10.1. CHROMOSOME PREPARATIONS

THE salivary glands of many Dipteran larvae consist of enlarged cells whose nuclei are also much increased in size. The chromosomes present duplicate themselves many times, each duplicate remaining alongside the original, forming what appears to be a giant chromosome. This consists, in fact, of bundles of duplicates (Plate XIII).

Biological requirements

Chironomus larvae
Drosophila larvae

The most convenient material to use is the salivary glands from *Drosophila* larvae. These larvae should be well fed, and grown at a temperature of 15°–18°C. When they are fully grown the larvae climb the sides of the glass container and remain motionless before pupating. These are the best stages to use.

Unfortunately *Drosophila* larvae, even when grown slowly are still relatively small and difficult to handle. *Chironomus* larvae are much larger than the former, they have larger salivary glands with larger chromosomes. The single disadvantage in their use is a small one, they have to be obtained by sifting through the mud from the bottom of ponds. They are a red in colour often with a green sheen, and they are from 10 to 20 mm in length.

Chemical reagents

(i) Insect saline (7·0g NaCl/litre solution)
(ii) Acetic orcein

Apparatus

Dissecting microscope or mounted lens
Microscope slides and cover slips

Microscope with green filter
Filter paper
Mounted needles

Dissecting out the salivary glands

The removal of the salivary glands is similar in both larvae. In the case of *Drosophila* a mounted lens or dissecting microscope will make the work easier. Place the larva on a slide in a few drops of insect saline. Anchor it by pressing down with a needle on the rear portion of the larva. Then using a second needle decapitate the insect by pulling the head away from the remainder. Besides a portion of the gut the salivary glands should emerge attached to the head. They appear as semi-transparent oval structures mixed with the body contents. If they have not emerged use the side of the second needle to squeeze the anterior portion of the body, stroking towards the cut end. This should dislodge the glands.

Clear away the debris: this is most conveniently done by moving the glands to a drop of saline on a clean slide. Never allow them to dry out.

Staining

Cover the salivary glands with acetic orcein which will both fix and stain the tissue. Leave for three to five minutes. Apply a coverslip and then carefully turn the slide over so the coverslip is on the under side. Place the slide in the centre of a filter paper which is on a perfectly flat surface. Press the slide vertically downwards. This will squash the tissue and spread out the chromosomes. The excess stain will be absorbed by the filter paper. If the preparation shows signs of drying out add a drop of stain to the edge of the coverslip. Examine under the high power of a microscope using a green filter.

10.2. THE GENETIC CONTROL OF ENZYME SYNTHESIS AND COMPLEMENTATION IN *Coprinus*

The organism used to demonstrate the link between genes and metabolic pathways is *Coprinus lagopus*, one of the ink-cap fungi. The metabolic pathway considered is the synthesis of the compound choline from ethanolamine (Fig. 84). Three basic steps are involved. The ethanolamine is first converted to monomethyl ethanolamine: this, by the addition of a second methyl group, is converted to dimethyl ethanolamine. The addition of a third methyl group results in the formation of choline. Each step in the pathway is controlled by a separate enzyme. The wild-type (WT) *Coprinus* will grow on a minimal medium of mineral salts, asparagine, glucose and thiamine. But two mutants, called Chol–1 and Chol–2, are only able to grow on this medium if choline is added. This means that they are unable to synthesise the choline from the compounds present in the minimal medium. To determine at which stage the synthesis of choline is being blocked the two mutants are inoculated on to several plates of minimal media each of which contains one intermediate compound from the choline pathway.

Neither mutant is able to grow on minimal medium containing ethanolamine or

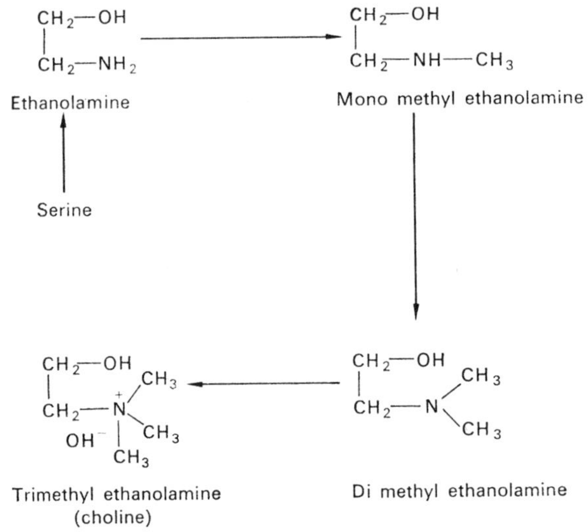

FIG. 84
The biosynthesis of choline

monomethyl ethanolamine. But growth of Chol–1 mutants is good on dimethyl ethanolamine and choline. This means that the Chol–1 mutants lack the enzyme necessary to convert monomethyl ethanolamine to dimethyl ethanolamine. The Chol–2 mutant is only able to grow on medium enriched with choline and thus must lack the enzyme which acts on dimethyl ethanolamine.

Complementation

The fact that mutation in both Chol–1 and Chol–2 strains blocks the choline pathway, might be due to gene mutation at the same locus. Proof that the genes concerned are at different loci can be obtained by complementation tests. If two compatible strains of the two mutants are plated together on to minimal medium, growth occurs, even though the mutants will not grow if plated separately. This growth is due to the fusion of the two mating strains of hyphae forming a dikaryon (Fig. 85). The dikaryotic hyphae contain the nuclei from the two compatible parents. These nuclei do not fuse but lie side by side and divide synchronously. The dikaryon is able to grow on minimal medium because the nuclei of the Chol–1 strain carry a wild-type gene at the Chol–2 locus: the nuclei of the Chol–2 strain likewise carry a wild-type gene at the Chol–1 locus. Each can therefore synthesise the enzyme missing from the other strain, in other words they complement each other (Fig. 85). If the two mutated genes were at the same locus no complementary wild-type gene would exist and the dikaryon would be unable to grow on the minimal medium. Together these two sections support the 'one gene–one enzyme' theory.

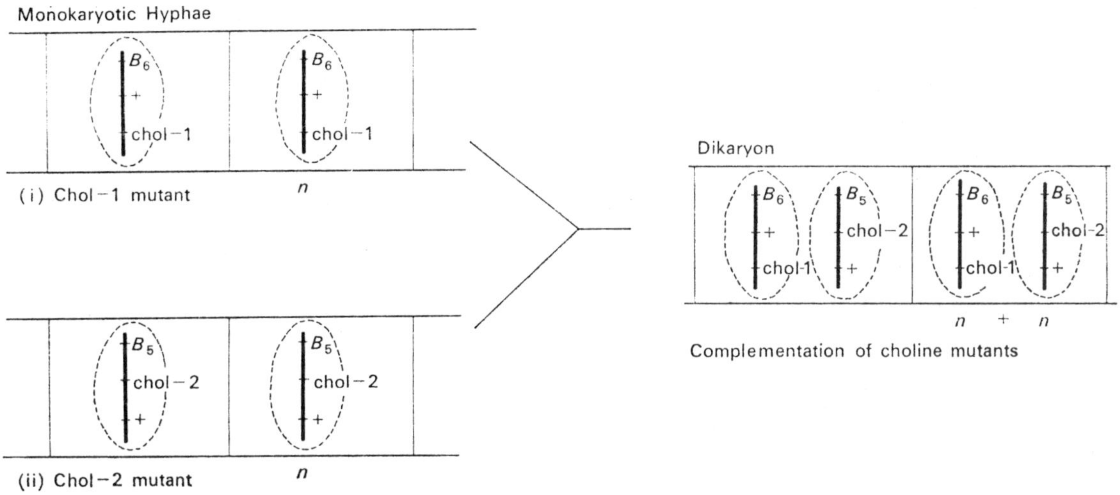

FIG. 85
Dikaryon formation and complementation in *Coprinus*

Compatibility

Two loci are concerned with mating in *Coprinus* and they are situated on different chromosomes. The loci are termed the *A* locus and the *B* locus and at each there are numerous alleles differentiated numerically, that is $A_1A_2A_3$, B_1 and so on. In order to produce a dikaryon and fruiting body the genes on both loci must be different, so that $A_1B_1 \times A_2B_2$ is a compatible mating whereas $A_1B_1 \times A_1B_2$ is incompatible and no dikaryon will be formed. The whole of the above work can be carried out as one experiment.

Biological requirements

Coprinus lagopus: wild type (any mating strain)
Chol–1 mutant, mating type A_6B_6
Chol–2 mutant, mating type A_5B_5

Chemical requirements

(i) Complete medium for *Coprinus*
(ii) Minimal medium

(iii) Minimal plus ethanolamine
(iv) Minimal plus monomethyl ethanolamine
(v) Minimal plus dimethylethanolamine
(vi) Minimal plus choline

Apparatus

Sterile Petri dishes
Inoculating loop
Wax pencil
Bunsen burner

Using sterile techniques (see page 170) pour out six plates, each containing one of the above media. Label each one. Using the wax pencil mark out the bottom of each dish as in Fig. 86. Plate out small portions of the relevant strains on to their positions in each of the six dishes. Incubate at 37°C for about two days.

For any further information on *Coprinus lagopus* and fungal genetics in general see *The Life History and Genetics of Coprinus lagopus Illustrated*, a practical introduction to biochemical genetics by G. E. Anderson, from which the above information was taken with the kind permission of the author.

Harris Biological Supplies, who publish the above book, also supply *Coprinus* cultures and the range of culture media.

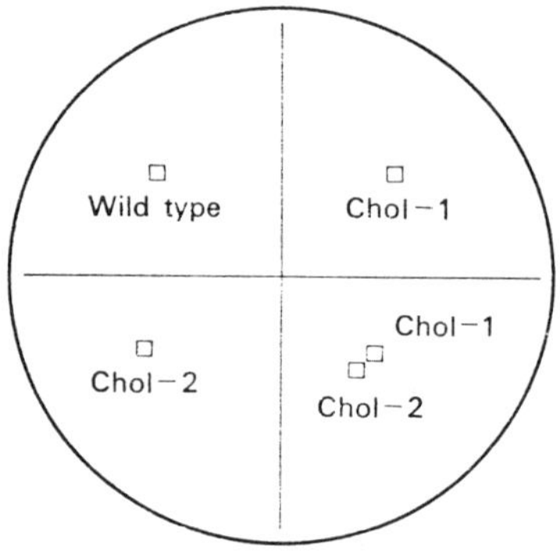

FIG. 86
Arrangement of *Coprinus* strains on culture plates

10.3. THE INDUCTION OF ENZYME SYNTHESIS

In Chapter 2 we discussed the Jacob-Monod experiment which showed that *Escherichia coli* grown on a medium containing lactose would begin producing enzymes (β galactosidase and permease) to deal with the lactose: the enzymes were not formed by bacteria grown in the absence of the lactose. This experiment illustrates one theory of how genes may be switched on, the presence of the substrate causing the DNA/RNA system to produce the necessary enzymes.

This induction of enzyme synthesis can also be demonstrated using seedlings grown in the presence or absence of nitrate. Normally plants taking in nitrates reduce this to nitrite. The nitrite is then converted to ammonia using reduced ferrodoxin, formed during photosynthesis, as the hydrogen donor. The ammonia then enters the metabolic pathways responsible for amino acid synthesis. The original reaction, that of the conversion of nitrate to nitrite, is brought about by the enzyme nitrate reductase. The synthesis of this enzyme is induced by the presence of nitrate ions in the surrounding medium. Seedlings grown in the absence of nitrate, or on an alternative source of nitrogen such as ammonium compounds, do not synthesise the enzyme.

Biological requirements

Seeds of soy-bean, barley, wheat, oats or radish.

Chemical reagents

(i) Propanol
(ii) Sulphanilic acid (a 1% solution in 3N HCl)
(iii) Naphthylethylene diamine dihydrochloride (*obtainable from BDH*) (0·02% solution in water, store in a dark bottle)
(iv) 0·IM sodium phosphate buffer pH 7·5
To make the buffer prepare two stock solutions:
(A) A solution of 35·7 g disodium hydrogen phosphate ($Na_2HPO_412H_2O$) per litre (0·IM solution)
(B) A solution of 15·6 g sodium dihydrogen phosphate ($NaH_2PO_42H_2O$) per litre (0·IM solution)
To obtain a buffer of pH 7·5 mix 350 ml (A) with 50 ml (B)
(v) 50 mM potassium nitrate solution
(Dissolve 2·2 g KNO_3 in the above buffer solution and make up to 400 ml)

Nutrient solutions

(vi) 50 mM potassium nitrate solution
(2·2 g KNO_3 made up to 400 ml solution using distilled water)
(vii) 50 mM ammonium sulphate solution
(2·7 g $(NH_4)_2SO_4$ made up to 400 ml solution using distilled water)

Apparatus

Petri dishes
Cotton wool
1 ml pipettes
100×25 mm specimen tubes
Test tubes
Aluminium foil

Germinate the seeds on damp cotton wool in three Petri dishes—use distilled water only. Treat the three groups of seedlings as follows:

(i) Feed with distilled water
(ii) Feed with about 5ml of 50 mM ammonium sulphate solution (vii)
(iii) Feed with about 5ml of 50 mM potassium nitrate solution (vi).

The next day remove sections of plant tissue to give 0·2–0·5g fresh weight and place them in a specimen tube together with 5ml of potassium nitrate in buffer solution, and 0·2ml propanol (this makes the tissues permeable). Repeat for the other two groups of seedlings and wrap the three specimen tubes in foil or place in the dark. If the plant tissue contains the enzyme nitrate reductase it will convert some of the surrounding nitrate into nitrite. By placing the tubes in the dark photosynthesis is inhibited and the conversion of the nitrite to ammonia prevented.

Leave the tubes for thirty minutes, then remove 1 ml of the solution, pipette into a test tube, add 1 ml of sulphanilic acid and mix. Then add 1 ml of naphthylethylene diamine and mix. Repeat this with the other two tubes. Leave the tubes for fifteen minutes. If nitrite is present a pink colour develops, the intensity of the colour is a measure of the amount of nitrite in the solution, which in turn is a measure of the amount of nitrate reductase in the tissue.

10.4. MITOSIS IN ROOT-TIP PREPARATIONS

Biological requirements

Crocus balansae	2n = 6
or *Allium cepa* (onion)	2n = 16
or *Allium sativum* (garlic)	2n = 16
or *Tradescantia paludosa*	2n = 12

Actively growing roots are required. The bulbs can be induced to form roots by suspending them over a beaker of water with the base of the bulb touching the water. The *Tradescantia* will supply root tips for most of the year if kept in a warm laboratory or greenhouse: to obtain the roots invert the pot and tap the rim to dislodge the root ball. Select the thickest actively growing roots with root hairs.

Chemical reagents

(i) Acetic alcohol fixative (1 part glacial acetic acid: 3 parts absolute alcohol, dehydrating alcohol mixtures will substitute for ethanol)
(ii) Acetic orcein (or propionic orcein)
(iii) N Hydrochloric acid

Apparatus

Microscope with green filter
Microscope slides and cover slips
Watch glasses
Mounted needles
Blunt forceps
Filter paper

The root tips can be used fresh but better results are obtained if they are fixed for a day or so in acetic alcohol.

Remove about 3mm of the tip of the root, place it in a watch glass and add 10 drops of acetic orcein and 1–2 drops of N hydrochloric acid. Holding the watch glass with the blunt forceps, warm it gently above a low flame, but do not allow it to boil. The acid macerates the cells allowing them to separate more easily and enabling the stain to penetrate. Continue this treatment for 3–4 minutes. Remove the tip on to a clean slide and add 2–3 drops of acetic orcein, then using mounted needles break up the root tip into smaller fragments. Apply a coverslip, and carefully turn the slide over. Place the slide in the centre of a filter paper on a flat surface. Press the slide vertically downwards to squash the tissue and spread the cells apart. The excess stain will be absorbed by the filter paper, but if the preparation shows signs of drying out add a drop of stain to the edge of the coverslip. Examine under the high power of the microscope using a green filter.

10.5. MEIOSIS IN LOCUST TESTIS PREPARATIONS

Biological requirements

Young adult locusts (male)

Chemical reagents

(i) Insect saline (0·7% sodium chloride solution)
(ii) Acetic orcein

Apparatus

Microscope with green filter
Microscope slides and coverslips
Dissecting instruments
Dissecting dish with black wax

Kill the insect by etherising it, and then decapitating it. Make a median longitudinal incision through the dorsal region of the abdomen. Place the animal in the dissecting dish and cover with insect saline: pin open the body wall to expose the contents of the abdomen. The testes lie in the anterior region of the abdomen, and are usually embedded in yellow fat bodies. Dissect out this material on to a slide and carefully search out the follicles from amongst the fat. The follicles of the testes are transparent or white and banana-shaped. Place two or three follicles on to a clean slide and using a thin scalpel smear them in a drop of acetic orcein. Then proceed as in 10.1.— Chromosome preparations.

10.6. MENDELIAN INHERITANCE IN *Drosophila*

Drosophila melanogaster is a small fly which occurs in all tropical and temperate countries feeding on the yeasts which grow in the juices of decayed fruit. In the laboratory it is fed on artificial medium and at 25°C gives a new generation every 12–14 days. The offspring are produced in large numbers. It has many kinds of hereditary variations which are easily recognised, and its somatic cells contain only four pairs of chromosomes. All of these features make it an ideal organism for genetic study.

Stock cultures should be reared in bottles the size of $\frac{1}{3}$ or $\frac{1}{2}$ pint milk bottles: experimental populations are reared in small tubes (100 × 25 mm) containing artificial medium.

Artificial medium

To make sufficient medium for ten stock bottles soak 25g porridge oats and 5g agar in 250ml water for twenty minutes. Then bring this mixture to the boil stirring continuously to prevent burning. When boiling add one tablespoon of black treacle and a pinch of Nipagin antimould. While it is still hot pour into the sterilised bottles. If no autoclave is available to sterilise the bottles, thoroughly wash them in hot water. Stopper the bottles with cotton wool or foam plastic plugs. When the medium is cool add a sprinkling of dried yeast, and a strip of filter paper. Bottles prepared in this way can be stored in a refrigerator until required, although care has to be taken to ensure that they do not dry out.

Life cycle

The duration of each stage is affected by several environmental factors, temperature being the most important. The eggs are laid on the surface of the medium and hatch in about twenty-four hours. The larvae tunnel their way through the medium while feeding and 'much tunnelling' indicates a successful culture. The fully grown larvae pupate on the strip of paper provided, or on the sides of the tube. The adults emerge about a fortnight after the eggs were laid, and they may live for up to a month.

Sexing the flies

Males may be recognised by their more rounded abdomens and by the more solid black pigmentation of the tip of the abdomen. The female's abdomen is larger and more obviously striped (Fig. 87). In newly emerged adults sexing is often difficult due to lack of pigmentation, but one certain method is to use a lens and examine the under-

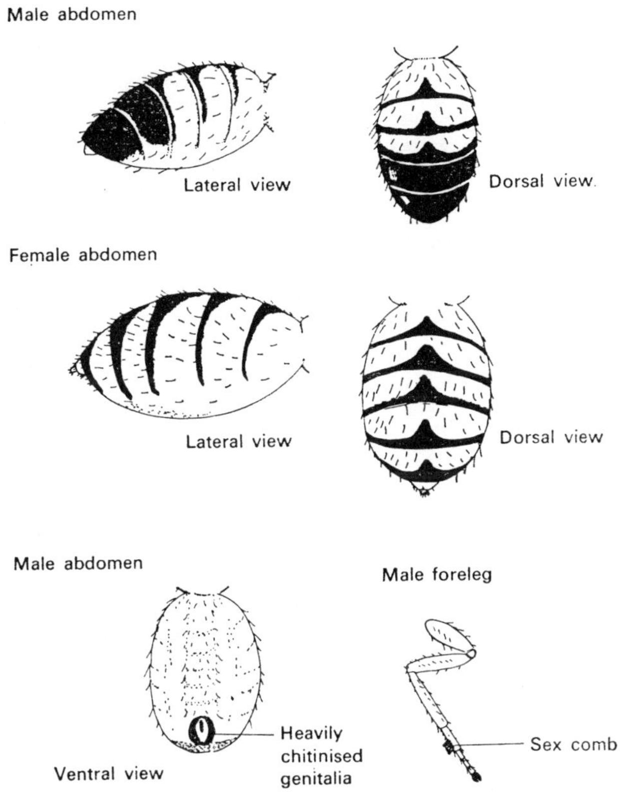

FIG. 87
The difference between male and female *Drosophila*

neath of the abdomen. In males the genitalia are heavily chitinised and easily recognisable as a brown circular structure on the posterior region of the ventral surface. The male fly also has a sex comb on the ventral surface of the fore-limbs.

Examining the flies

Before the flies can be examined and sorted they have to be anaesthetised using ether vapour. Tap the culture tube gently but rapidly against your hand, forcing the flies to the bottom of the tube. Quickly remove the cotton wool plug and invert the tube, mouth to mouth, over the etherising tube (Fig. 88). Holding the two tubes firmly together tap the etherising tube against your hand to dislodge the flies into this tube. Quickly stopper the etherising tube with its bung and replace the cotton wool in the culture tube. Add 3–4 drops of ether to the cotton wool in the top of the etherising tube.

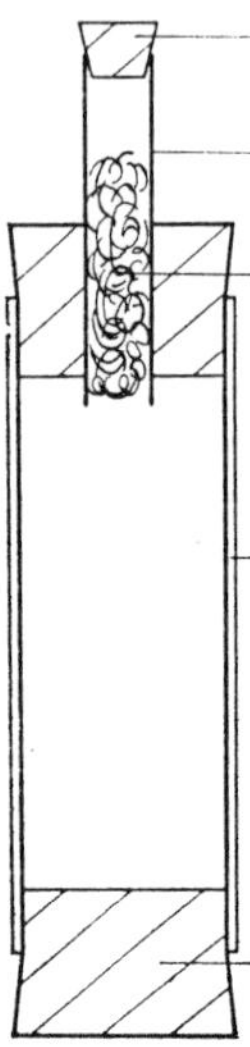

FIG. 88
Drosophila etheriser

As soon as the last fly stops moving remove the bung and empty the flies on to a white tile. Do not over-etherise or the flies will die. Death is indicated by a bunching of the legs and wings folded over the back. The flies are easily injured and a small brush should be used when moving them about on the tile. Etherised flies will recover in about five minutes. They may be re-etherised on the tile by inverting a filter funnel over them and placing a few drops of ether on a plug of cotton wool in the neck of the funnel. To transfer flies from the stock bottle to the etheriser, replace the small tube in the bung of the etheriser with a 5cm filter funnel.

Setting up experimental crosses

About ten hours after emergence the flies will begin to mate. During mating the females will store considerable amounts of sperm: fertilisation occurs later when the eggs are about to be laid. When making experimental crosses it is essential that the females are *virgin* as stored sperm would interfere with the results. Virgin flies are obtained from a given culture by removing *all* flies from the culture; any flies emerging within the following ten hours will be virgin. When setting up an experimental cross, put three virgin females with six males per culture; any extra could result in too heavy a population of larvae. When anaesthetised flies are placed into a culture bottle, always leave the bottle on its side until the flies have recovered, thus eliminating the possibility that the

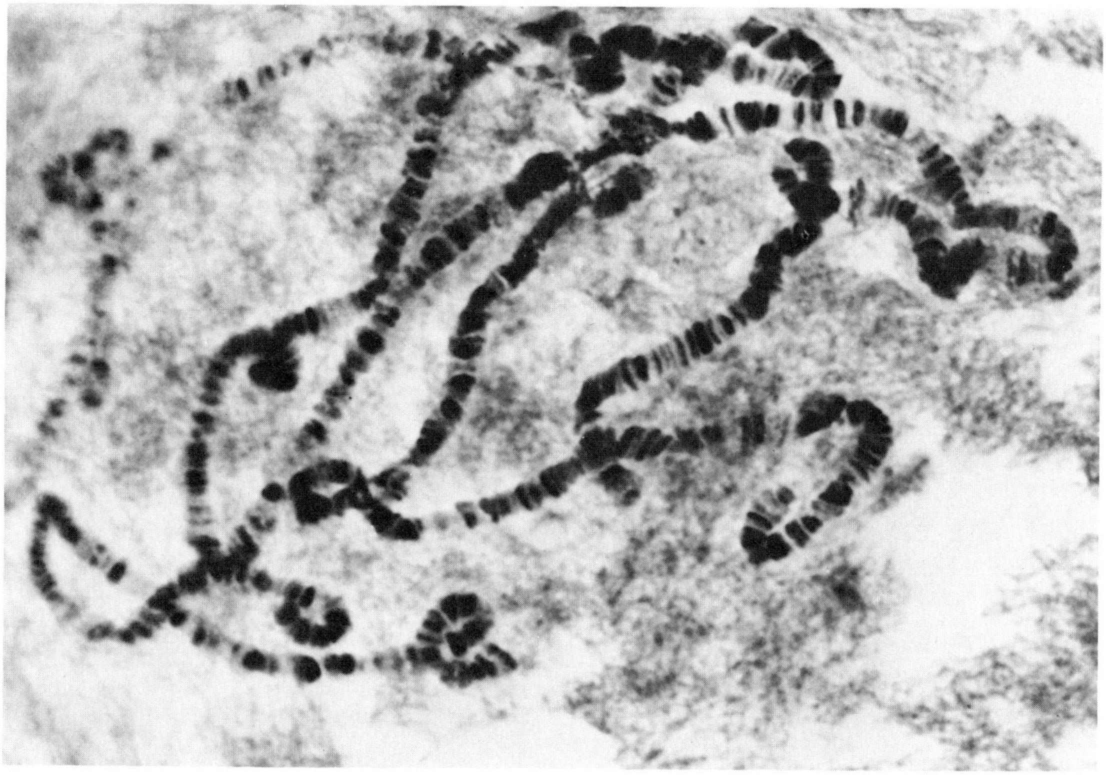

PLATE XII
Drosophila melanogaster. Salivary gland chromosomes (giant chromosomes).

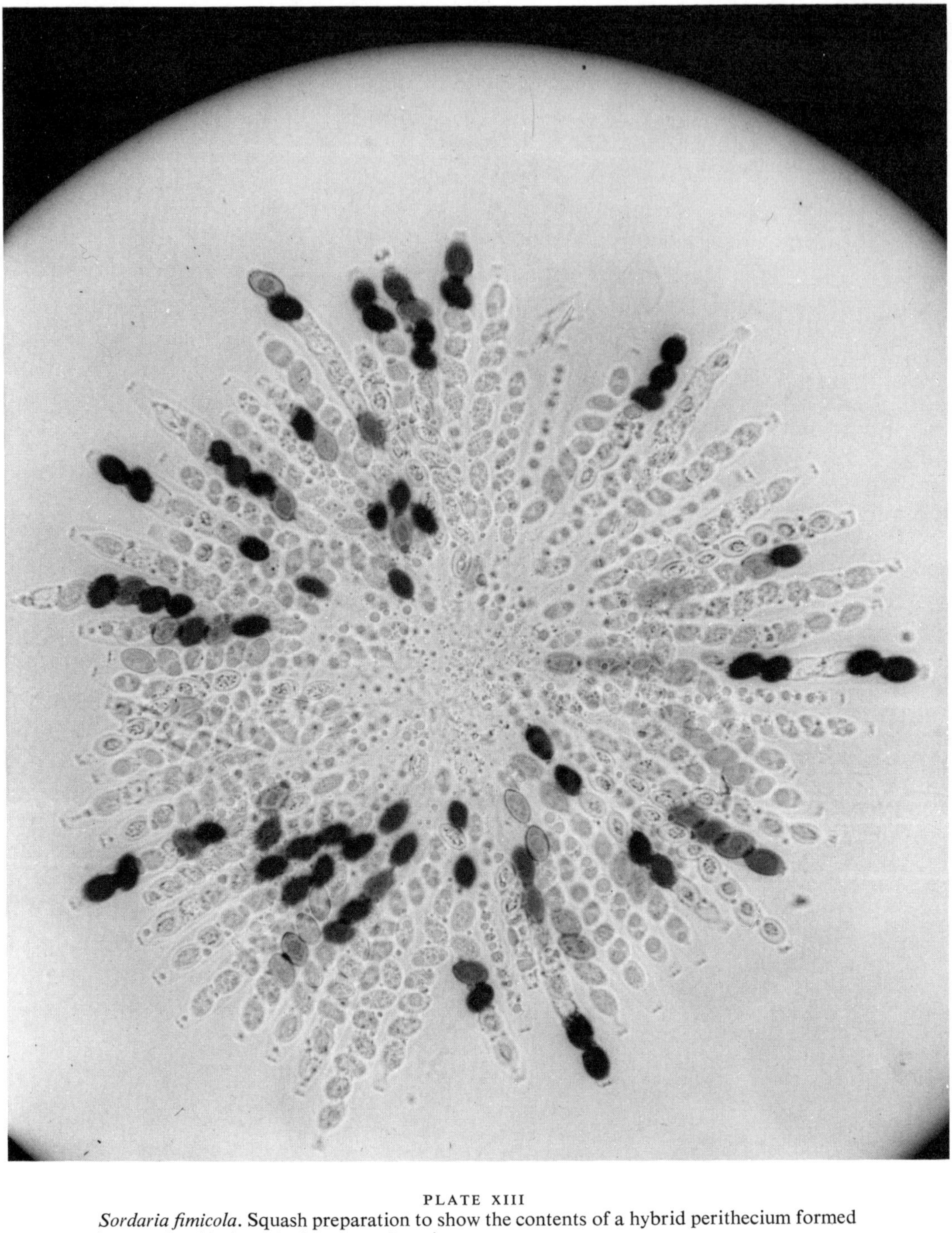

PLATE XIII

Sordaria fimicola. Squash preparation to show the contents of a hybrid perithecium formed by crossing black and white spored strains.

flies will become embedded in the medium and thus die. Label the culture tube with the type of cross and the date it was set up.

Suggested crosses

(i) *Monohybrid cross:*
  Wild type × ebony body
  Wild type × vestigial wing

(ii) *Dihybrid cross; with independent assortment*
  Ebony body × vestigial wing

(iii) *Autosomal linkage*
  Ebony body × curled wing

(iv) *Sex linkage*
  Wild type × white eye

It is not necessary to select virgin females from the *F1* population where these flies are going to be inbred to produce the *F2*. To obtain the *F2* generation, anaesthetise the *F1* flies, examine and score the phenotypes, then select six males and three females and place these in a fresh culture tube.

To show the effect of a testcross mate *F1* flies from the monohybrid cross with the homozygous recessive strain.

When scoring the results of the *F2* generations count the flies emerging on several consecutive days as some mutants have a delayed development.

Unwanted excess flies should be killed by placing them in a 'morgue' (methylated spirit in a wide-mouthed specimen jar). This reduces the risk of escaped flies interfering with the stock cultures.

10.7. 3:1 SEGREGATION IN TOBACCO (*Nicotiana tabacum*)

Gene mutation in tobacco has given rise to a strain of plants which are incapable of forming chlorophyll. Plants homozygous for this mutated gene are albino, and because they are incapable of photosynthesising the seedlings are short-lived. If plants which are heterozygous for this gene are selfed the seed produced can be expected to give rise to both normal and albino seedlings in the ratio of 3:1. This type of seed can be obtained from suppliers, and if sown on fine sterile compost in Petri dishes will produce seedlings in about a fortnight at 20°–25°C. The seed is very fine and should be mixed with find sand to get even dispersal. These seedlings should be grown in the light. The numbers of normal and albino seedlings can be scored and then the chi-squared test (Chapter 5) applied to the results.

If a similar experiment is set up but this time grown in the dark, the effect of an environmental factor on the expression of the normal gene can be demonstrated (Chapter 6).

A similar mutant is also available in barley.

10.8. CROSSING OVER IN *Sordaria fimicola*

The theory behind this technique was described in Chapter 5.

Biological requirements

Sordaria fimicola black spored (wild type) and white spored mutant

Chemical requirements

(i) Lactophenol
(ii) Malt agar

This can be made by adding 15g agar to 1000ml water and allowing the mixture to stand for thirty minutes. Then bring to the boil to dissolve the agar. Add 20g malt extract dissolved in a little hot water. Pour into 'medical flats' or McCartney bottles, replace the caps *loosely* and autoclave at 15lb pressure for twenty minutes.

Apparatus

Sterile Petri dishes
Inoculating loop
Bunsen burner
Wax pencil

Pouring plates

Bottles of the medium should be heated in a saucepan of boiling water to melt the agar. They are then allowed to cool to 45°–50°C. Holding the bottle in the right hand unscrew the cap by holding it between the little finger and the palm of the left hand—do not put the cap down. Flame the mouth of the bottle, and lifting the lid of the Petri dish far enough to allow the introduction of the neck of the bottle pour the medium into the dish. Replace the lid of the Petri dish, then screw on the cap of the bottle. Gently rotate the Petri dish to distribute the agar evenly over the bottom, allow to cool.

Inoculating the plate

Flame the inoculating loop (nichrome wire bent into a loop) until it is red hot, then allow it to cool briefly, without putting it down! Unscrew the cap from the culture bottle, holding the cap in the hooked little finger of the right hand, and the bottle in the left hand. Flame the mouth of the bottle. Using the inoculating loop remove a small piece of fungal growth, if necessary with a small portion of the agar, and plate it on the medium in the Petri dish. Only lift the lid of the dish sufficiently to allow the insertion of the loop. Flame the mouth of the culture tube and replace the cap. Flame the inoculating loop.

Sordaria fimicola preparation

The two strains of *Sordaria* should be plated at opposite sides of the Petri dish about 2cm from the edge. The dish should be labelled and then incubated at 25°C. *Sordaria* is homothallic and therefore self-matings will occur; however, where the two strains meet out-mating will give rise to perithecia containing both colours of ascopore. The perithecia appear after about a week, but the spores take another week or so to develop their full colour and immature black ascospores look very similar to the white form.

Examining the perithecia

Using the inoculating loop, remove a perithecium on to a slide, take with it as little agar as possible. Add a drop of lactophenol and apply a coverslip. Then using the blunt end of the handle of the inoculating loop, press gently on the coverslip directly over the perithecium. This should disrupt the perithecium and the sausage-shaped asci should emerge. The amount of pressure required to display the asci at their best is a matter of experience and luck. Examine the preparation for non-cross-over forms (4 black:4 white) and for asci where crossing over has occurred (see Chapter 5 and Plate XIV).

10.9. POLYPLOIDY IN *Tradescantia*

The phenomenon of polyploidy can be demonstrated using *Tradescantia* plants. These plants are available with triploid and tetraploid chromosome counts, as well as the normal diploid condition.

T. brevicaulis $2n=12$
T. virginiana $4n=24$
or
T. ohioensis $4n=24$

The plants show distinct phenotypic characters (see Plate VII) and examination of the root tip squash shows the increase in chromosome numbers, although it is very difficult to count the actual numbers present.

10.10. THE EFFECTS OF X-RAY IRRADIATION IN BARLEY

The effects of irradiation can be demonstrated by growing barley seeds which have been exposed to varying doses of irradiation. The plants formed from such seeds are often stunted and have decreased fertility, mutations which do occasionally appear at this stage are mainly somatic mutations. Genetic mutations which have occurred are more likely to show their presence in the progeny of these irradiated plants. The plants should be self-pollinated and the seeds grown. Only a very small percentage of the progeny will show any form of recognisable mutation.

For further information on irradiation of seeds see the *Radioisotopes Review Sheet*—'The Production of mutations by gamma irradiation of seeds' from the A.E.R.E. Sets of irradiated seeds are available from suppliers.

10.11. GENES IN POPULATIONS

As an exercise in population genetics take a random sample of one hundred pupils of the same age and score the following phenotypic characters:

(i) Ability to taste P.T.C. or not
(dissolve 1·3 g phenylthiocarbamide in boiling water and make up to 1 litre—soak pieces of filter paper in this solution, dry them and store in a plastic bag).

(ii) Whether the bottom of the ear is attached to the side of the head or whether there is a definite lobe.

(iii) Whether the person is able to roll the edges of the tongue together or not.

Given the frequency of the phenotypes, use the Hardy-Weinberg equation to calculate the frequency of the different genes concerned.

Glossary

Alleles – Alternative forms of a single gene.
Allopolyploid – An organism with more than two sets of chromosomes derived from different species by hybridisation.
Angstrom Unit (Å) – 1×10^{-8} cm.
Anisogamy – The formation of two clearly distinguishable types of gamete.
Asexual reproduction – Reproduction involving only one parent and without gametic fusion.
Autopolyploid – An organism having more than two sets of homologous chromosomes.
Autosomes – Chromosomes which are not involved in determining sex.
Bivalent – The structure formed by the association of the members of a pair of homologous chromosomes during meiosis I.
Centriole – Structures visible particularly in animal cells, associated with the formation of the spindle in cell division.
Centromere – A specialised region of the chromosome at which the chromatids are joined.
Chiasma – A point at which members of a pair of homologous chromosomes are apparently joined during prophase and metaphase of meiosis I.
Chromatids – The two identical parts of a chromosome present after replication. Visible during prophase and metaphase of mitosis and meiosis.
Chromonemata – The fine strands from which chromosomes are built up. Each consists of a combination of DNA and protein.
Chromosome – Thread-like structures found in the nuclei of cells, carrying the genetic information.
Cleistogamy – Self-fertilisation in flowering plants taking place before the flower structures open.
Crossing-over – A process by which lengths of chromosome are exchanged between members of a homologous chromosome pair during meiosis I. Chiasmata represent points of crossing-over.
Deficiency – Inactivation or absence of a chromosomal segment or gene.
Diploidy – The presence within the nucleus of two homologous sets of chromosomes.
DNA – Deoxyribonucleic acid.
Dominance – The phenomenon by which only one of a pair of contrasting alleles present in the genotype is expressed in the phenotype.
Duplication – The presence of a chromosome fragment in addition to the normal complement.
Endoplasmic reticulum – A unit membrane network within the cell, linking nucleus, cytoplasm and cell membrane.
Epistasis – Dominance of a gene over another, non-allelomorphic, gene.
Equator – The central plane of the spindle on which the chromosomes lie during metaphase of cell division.
Eugenics – The application of artificial selection to human populations in order to increase the frequency of favourable genes and decrease the frequency of harmful genes.
F1 – Denotes the first filial generation. Successive generations are shown as *F2*, *F3*, etc.
Fertilisation – See *syngamy*.

Gametes – Specialised reproductive cells formed by the parental organisms during sexual reproduction.
Gametogenesis – The formation and maturation of the gametes.
Gene – A length of DNA with a particular sequence of organic bases coding for the formation of a protein.
Genotype – The total genetic constitution of an organism.
Haploidy – The presence within the nucleus of a single set of chromosomes.
Heterosis – The increase in viability and vigour displayed by heterozygous organisms.
Heterozygote – An organism or cell containing two contrasting alleles of a single gene.
Homologous pair – Two similar chromosomes present in a diploid cell, one being derived from each parent.
Homozygote – An organism or cell containing two identical alleles of the gene.
Inversion – The reversal of a segment of a chromosome within the chromosome as a whole.
Isogamy – The formation of only a single type of gamete.
Karyotype – The chromosome set characteristic of a given species.
Linkage – The inheritance of separate genetic characters in a non-random manner. Caused by the presence of separate genes on the same chromosome.
Locus – The point on a chromosome at which a gene is situated.
Meiosis – A method of cell division giving daughter cells with half the number of chromosomes of the parent cell.
Micron (μ) – 1×10^{-3} mm.
Mitosis – A method of cell division giving daughter cells with chromosomes identical to those of the parent cell.
Mutagen – An agent capable of inducing mutation, for example ionising radiations.
Mutation – A change in the genetic material involving the sequence of organic bases in the DNA, part of a chromosome, whole chromosomes or complete sets of chromosomes.
Non-disjunction – Failure of homologous pair of chromosomes to separate during meiosis.
Nucleolus – A darkly staining area within the nucleus visible during interphase. It is often associated with a single chromosome.
Nucleotide – The basic unit from which DNA and RNA are made up. It contains sugar, phosphate, and organic base groups.
Phenotype – The sum total of an organism's characteristics.
Pleiotropy – Where a single gene has multiple effects on the phenotype.
Polyploid – Having duplicate sets of chromosomes, for example triploid, tetraploid and so on with three and four times the normal haploid number of chromosomes.
Polysomes – A collection of ribosomes, probably associated with a single molecule of mRNA.
Protandry – The condition in hermaphrodite organisms where the male gametes mature and are dispersed before the female gametes develop.
Protogyny – The condition in hermaphrodite organisms where the female gametes mature before the male gametes.
Ribosomes – Small granular structures attached to parts of the endoplasmic reticulum. The site of protein synthesis.
RNA – Ribonucleic acid.
Sex chromosomes – The chromosomes whose presence, absence or form may determine sex.
Sex-linked inheritance – The inheritance of genes carried on the sex chromosomes.
Sexual reproduction – Reproduction involving the fusion of two gametes, usually from separate parental organisms.
Spindle – A fibrous structure visible within the cell during division, apparently controlling chromosome movement.
Synapsis – The exact pairing of homologous chromosomes during the early stages of meiosis.
Syngamy – The fusion of the gametes in sexual reproduction.
Testcross – A method of determining whether a dominant allele is present in the homozygous or heterozygous state by crossing it with an organism which is homozygous recessive.

Transformation – The transfer of genetic information in the form of DNA, between related bacteria.
Translocation – The change in the position of a chromosome segment to another part of the same, or a different, chromosome.
Trisomic – Having the normal chromosome complement plus an additional single chromosome.
Zygote – The cell formed during sexual reproduction by the fusion of the two gametes.

Index